Getting Energy Prices Right
From Principle to Practice

Ian Parry, Dirk Heine, Eliza Lis, and Shanjun Li

INTERNATIONAL MONETARY FUND

© 2014 International Monetary Fund

Cataloging-in-Publication Data
Joint Bank-Fund Library

Getting energy prices right : from principle to practice / Ian Parry, Dirk Heine, Eliza Lis, and Shanjun Li. – Washington, D.C. : International Monetary Fund, c2014.
       pages ; cm

Includes bibliographical references.
ISBN: 978-1-48438-857-0

1. Power resources—Prices.   I. Parry, Ian.   II. Heine, Dirk.
III. Lis, Elisa.   IV. Li, Shanjun.   V. International Monetary Fund.

HD9502.A2 G48 2014

Disclaimer: The views expressed in this book are those of the authors and should not be reported as or attributed to the International Monetary Fund, its Executive Board, or the governments of any of its member countries.

Please send orders to:
International Monetary Fund, Publication Services
P.O. Box 92780, Washington, DC 20090, U.S.A.
Tel.: (202) 623-7430 Fax: (202) 623-7201
E-mail: publications@imf.org
Internet: www.elibrary.imf.org
www.imfbookstore.org

Dedication

This book is dedicated to Gary Becker, in appreciation of his inspiring teaching and long-time support of my work on environmental taxes; as my thesis chairman in the early 1990s, he encouraged me to model carbon taxes. Gary had agreed to write an endorsement for this book before his untimely death, his main reservation being the risk that environmental tax revenues might be overused for low-value spending.

Ian Parry

Contents

Foreword ix
Acknowledgments xi
Abbreviations xiii

1	**Summary for Policymakers**	1
2	**Energy Systems, Environmental Problems, and Current Fiscal Policy: A Quick Look**	11
	Overview of Energy Systems	11
	Environmental Side Effects	14
	Fiscal Policies Currently Affecting Energy and Transportation	24
3	**Rationale for, and Design of, Fiscal Policy to "Get Energy Prices Right"**	31
	Policy Instrument Choice for Environmental Protection	31
	Further Design Issues	44
	Summary	60
4	**Measuring Pollution Damage from Fuel Use**	65
	CO_2 Damage	65
	Local Air Pollution Damage	67
	Summary	90
5	**Measuring Nonpollution Externalities from Motor Vehicles**	101
	Congestion Costs	101
	Accident Costs	111
	Road Damage Costs	115
	Summary	117
6	**The Right Energy Taxes and Their Impacts**	131
	Corrective Tax Estimates	131
	Impacts	142
	Summary	145
7	**Concluding Thoughts**	165

Glossary 167
Index 173

Tables

4.1.	Examples of Mortality Risk Valuations Used in Previous Government Studies	81
5.1.	City-Level Travel Delays and Other Characteristics, Region Average, 1995	103
5.2.	Country-Level Travel Delays and Other Characteristics, Region Average, 2007	104
5.3.	Reviews of Empirical Literature on the Value of Travel Time (VOT)	107

Figures

1.1.	Corrective Fuel Taxes to Reflect Environmental Costs, Selected Countries, 2010	6
1.2.	Impacts of Fuel Tax Reform, Selected Countries, 2010	7
2.1.	Primary Energy Consumption per Capita, Selected Countries, 2010	12
2.2.	Electricity Consumption per Capita, Selected Countries, 2010	12
2.3.	Motor Vehicle Ownership Rates, Selected Countries, 2010	13
2.4.	Share of Final Energy Use by Fuel Type, Selected Countries, 2010	14
2.5.	Carbon Dioxide (CO_2) Emissions per Capita, Selected Countries, 2010	15
2.6.	Urban Population, Selected Countries, 2010	15
2.7.	Projected Global Energy-Related CO_2 Emissions	18
2.8.	Projected Long-Term Warming above Pre-Industrial Temperatures from Stabilization at Different Greenhouse Gas Concentrations	19
2.9.	Air Pollution Concentrations, Selected Countries, 2010	21
2.10.	Air Pollution Deaths by Region, 2010	22
2.11.	Vehicles and Road Capacity, Selected Countries, 2007	23
2.12.	Road Deaths, Selected Countries, 2010	24
2.13.	Revenue from Environment-Related Taxes as Percent of Total Revenue in OECD Countries, 2010	25
2.14.	Excise Tax Rates on Motor Fuels, 2010	26
2.15.	Subsidies for Fossil Fuel Energy by Region and Fuel Type, 2011	27
3.1.	Illustrated Sources of Fossil Fuel CO_2 Reductions under Different Policies	35
3.3.1.	Shape of the Air Pollution Damage Function	39
3.2.	Price Experience in the European Union Emissions Trading System	43
3.3.	Distributional Incidence of Energy Subsidies	58
4.1.	Baseline Mortality Rates for Illnesses Whose Prevalence Is Aggravated by Pollution, Selected Regions, 2010	75
4.2.	Value of Mortality Risk, Selected Countries, 2010	81
4.3.	Damage from Coal Plant Sulfur Dioxide (SO_2) Emissions, Selected Countries, 2010	82
4.4.	Damage from Coal Plant Sulfur Dioxide (SO_2) Emissions, All Countries, 2010	84

4.5.	Damage from Ground-Level Nitrogen Oxide (NO_x) Emissions, Selected Countries, 2010	85
4.6.	Estimated SO_2 Damage Relative to China Using the Intake Fraction Approach, 2010	86
4.7.	Estimated SO_2 Damage Relative to China Using the TM5-FASST Model, 2010	87
4.8.	SO_2 Emissions Rates at Coal Plants, 2010	90
5.1.	Value of Travel Time, Selected Countries, 2010	108
5.2.	Congestion Costs Imposed on Others per Car-Kilometer, Selected Countries, 2010	109
5.3.	Congestion Costs Imposed on Others per Car-Kilometer, All Countries, 2010	110
5.4.	External Accident Costs per Vehicle-Kilometer, Selected Countries, 2010	115
5.5.	External Accident Costs per Kilometer Driven, All Countries, 2010	116
6.1.	Corrective Coal Tax Estimates, Selected Countries, 2010	132
6.2	Corrective Taxes for Air Pollution at Coal Plants with and without Control Technologies, Selected Countries, 2010	133
6.3.	Breakdown of Air Pollution Damages from Coal by Emissions Type, Selected Countries, 2010	134
6.4.	Corrective Coal Tax Estimates, All Countries, 2010	135
6.5.	Corrective Coal Tax Estimates with Uniform Mortality Values, Selected Countries, 2010	136
6.6.	Corrective Natural Gas Tax Estimates for Power Plants, Selected Countries, 2010	137
6.7.	Corrective Natural Gas Tax Estimates for Power Plants, All Countries, 2010	138
6.8.	Corrective Gasoline Tax Estimates, Selected Countries, 2010	139
6.9.	Corrective Gasoline Tax Estimates, All Countries, 2010	140
6.10.	Corrective Diesel Tax Estimates, Selected Countries, 2010	141
6.11.	Potential Revenue from Corrective Fuel Taxes, Selected Countries, 2010	143
6.12.	Reduction in Pollution-Related Deaths from Corrective Fuel Taxes, Selected Countries, 2010	145
6.13.	Reduction in Energy-Related CO_2 Emissions from Corrective Fuel Taxes, Selected Countries, 2010	146

Boxes

2.1.	Broader Environmental Effects beyond the Study Scope	16
3.1.	Environmental Effectiveness of Alternative Instruments: Further Examples	33
3.2.	Defining Economic Costs	37
3.3.	Shape of the Air Pollution Damage Function	39
3.4.	Coverage of Energy Products under the Value-Added Tax (VAT)	40

3.5.	Environmental Tax Shifting in Practice	41
3.6.	Unintended Consequences and Market Price Distortions	45
3.7.	Examples of Distance-Based Charging for Vehicles	49
3.8.	Reconciling Fiscal and Environmental Objectives in Vehicle Taxation	52
3.9.	Pay-as-You-Drive Auto Insurance	53
3.10.	The Energy Paradox Controversy	56
4.1.	Intake Fractions: Some Technicalities	69
4.2.	The Human Capital Approach	77
4.3.	Determinants Other than Income of Mortality Risk Valuation	78
4.4.	Emissions Factors from the GAINS Model	88
5.1.	Broader Costs of Congestion	102

Appendix Tables

4.1.1.	Country Classifications for Baseline, Pollution-Related Mortality Rates	91
4.2.1.	Damage from Local Air Pollution, All Countries, $/ton of Emissions, 2010	92
5.2.1.	Cities in the City-Level Database (Used to Extrapolate Congestion Costs)	119
5.3.1.	Regression Results for City-Level Average Delay	121
5.3.2.	Regression Results for Kilometers Driven per Car	122
5.4.1.	Ratio of Congestion Cost with Multiple Vehicles Relative to Costs when Cars are the Only Vehicle	123
6.2.1.	Corrective Fuel Tax Estimates, All Countries, 2010	149
6.2.2.	Fiscal Impacts of Tax Reform, All Countries, 2010	153
6.2.3.	Health and Environmental Impacts of Tax Reform, All Countries, 2010	157
6.2.4.	Estimates of Current Fuel Excise Taxes, All Countries, 2010	161

Foreword

The enormous improvement in global living standards over the past 100 years could not have happened without the energy derived from the world's vast deposits of fossil fuels. Yet, the intense use of energy derived from such sources has brought side effects that pose major social, political, and economic challenges. The imperative is now to find ways to diversify and reduce the use of energy while continuing to eliminate poverty and promote inclusive growth. Within its mandate, the Fund is playing a part in this discussion.

Indeed, energy policies are not new territory for the Fund. We have been emphasizing the large fiscal benefits from removing harmful fuel subsidies for many years, primarily with an eye to saving money for the taxpayer. However, as the external effects of energy use have reached a macrocritical level—whether from environmental degradation, higher food prices, or the threat of climate change— the Fund's advice has also evolved.

Policymakers face a wide array of options for meeting energy challenges. Given the powerful incentive effect that prices have on economic behavior, the application of basic tax principles is critical. "Getting prices right" means that taxes on fossil fuels should be set at a level such that energy prices reflect their associated environmental side effects.

This economic principle is widely accepted; however, putting it into practice is by itself an intellectual challenge. Here lies the unique contribution of this book. After attempting to quantify the environmental impact of energy use, the authors calibrate, on a country-by-country basis, a system of fuel taxes that balances environmental benefits against economic costs. Where data allow, and with proper caveats, the book estimates appropriate fuel taxes for over 150 countries, and provides a framework for refining these estimates further.

The results confirm that many countries—advanced, emerging, and developing—are only at base camp with regard to getting energy prices right. Importantly, the results also show local air pollution damages, congestion costs, and revenue potential (e.g., in lieu of other taxes) are mostly large enough to warrant higher fuel taxes, even leaving climate concerns aside.

The tools and insights provided here should help us rise to the challenge that the pursuit of more effective energy pricing poses—and to recognize too the opportunities for more sustained, robust, and responsible growth that it presents.

Christine Lagarde
Managing Director
International Monetary Fund

Acknowledgments

We are indebted to many people for their help in the preparation of this report.

Paul Johnson and Michael Toman provided many thoughtful comments on the case for, and design of, environmental fiscal instruments in Chapter 3.

Maureen Cropper was especially helpful in suggesting and developing the intake fraction approach in Chapter 4. Nicholas Muller provided valuable feedback, ran the simulations using the TM5-FASST tool, and calibrated various relationships between pollution exposure and mortality risks. Fabian Wagner put great effort into compiling emissions factors for different fuels and different countries using the IIASA model. Ronnie Burnett provided valuable information about findings from the Global Burden of Disease Project on baseline rates of pollution-related mortality by region and evidence on how these rates respond to pollution exposure. Alan Krupnick and Neal Fann also provided many useful comments on the methodology and data sources.

Robin Lindsay and Erik Verhoef also offered many useful suggestions for improving the assessment and discussion of side effects from motor vehicles in Chapter 5. Jianwei Xing provided first-rate research assistance for the estimation of traffic congestion costs.

Saad Alshahrani helped with the estimation of the impacts of fuel tax reform. Martin Petri and Nidhi Kalra provided thoughtful perspectives on the presentation of the results and how they might be made useful for policymakers.

Kelsey Moser helped to produce many of the figures in the report, Louis Sears helped with baseline energy data, and Sherrie Barnes, Maria Delariarte, Maura Ehmer, Mary Fisk, and Madina Thiam assisted in the final preparation of the manuscript.

Participants at an IMF workshop at which an interim version of the report was presented also provided very helpful comments. The workshop participants (aside from those acknowledged above) included Dallas Burtraw, Martina Bosi, Ben Clements, David Coady, David Evans, Marianne Fay, Elizabeth Kopits, Richard Morgenstern, Adele Morris, John Norregaard, Wallace Oates, Jon Strand, Suphachol Suphachalasai, and Ann Wolverton. Maura Ehmer and Pierre Albert helped to organize the workshop.

Glenn Gottselig, Michael Keen, Ruud de Mooij, and Vicki Perry provided especially valuable guidance and suggestions on numerous occasions during the preparation of the report.

Abbreviations

°C	degrees centigrade
CO_2	carbon dioxide
ETS	emissions trading system
EU	European Union
FASST	Fast Scenario Screening Tool
GHG	greenhouse gas
GJ	gigajoule
IPCC	Intergovernmental Panel on Climate Change
NO_x	nitrogen oxide
PAYD	pay as you drive
PM	particulate matter
$PM_{2.5}$	particulate matter with diameter up to 2.5 micrometers
ppm	parts per million
R&D	research and development
SCC	social cost of carbon
SO_2	sulfur dioxide
U.K.	United Kingdom
U.S.	United States
VOT	value of travel time

CHAPTER 1

Summary for Policymakers

Many energy prices in many countries are wrong. They are set at levels that do not reflect environmental damage, notably global warming, air pollution, and various side effects of motor vehicle use. In so doing, many countries raise too much revenue from direct taxes on work effort and capital accumulation and too little from taxes on energy use.

This book is about getting energy prices right. The principle that fiscal instruments must be center stage in "correcting" the major environmental side effects of energy use is well established. This volume aims to help put this principle into practice by setting out a practicable methodology and associated tools for determining the right price. The book provides estimates, data permitting, for 156 countries of the taxes on coal, natural gas, gasoline, and diesel needed to reflect environmental costs. Underpinning the policy recommendations is the notion that taxation (or tax-like instruments) can influence behavior; in much the same way that taxes on cigarettes discourage their overuse, appropriate taxes can discourage overuse of environmentally harmful energy sources.

BACKGROUND

Energy use is a critical ingredient in industrial and commercial production, and in final consumption, but it can also result in excessive environmental and other side effects, with potentially sizable costs to the economy. For example,

- If left unchecked, atmospheric concentrations of carbon dioxide (CO_2) and other greenhouse gases (GHGs) are expected to raise global temperatures by about 3–4°C by the end of the century (IPCC, 2013). Temperature changes of this magnitude are large by historical standards and pose considerable risks.
- Outdoor air pollution, primarily from fossil fuel combustion, causes more than 3 million premature deaths a year worldwide, costing about 1 percent of GDP for the United States and almost 4 percent for China (National Research Council, 2009; World Bank and State Environment Protection Agency of China, 2007; World Health Organization, 2013).
- Motor vehicle use leads to crowded roads, accidental death, and injuries. Drivers in the London rush hour, for example, impose estimated costs on others that are equivalent to about US$10 per liter ($38 per gallon) of the fuel they use through their contribution to traffic congestion, and traffic accidents cause an estimated 1.2 million deaths worldwide (Parry and Small, 2009; World Health Organization, 2013).

The Need for Fiscal Policies

Given the seriousness of the problems associated with fuel use, addressing them with carefully designed policy instruments is critical. Ideally, these policies should do the following:

- Be *effective*—exploit all opportunities for reducing environmental harm and mobilizing private investment in clean technologies
- Be *cost-effective*—achieve environmental objectives at lowest cost to the economy
- Strike the right balance between the *benefits* and the *costs* of environmental improvement for the economy, thereby maximizing the net benefits.

All three features are important for balancing trade-offs between environmental protection and economic growth and enhancing prospects for sustaining and scaling up efficient policy. *Fiscal instruments*—environmental taxes or similar instruments (primarily emissions trading systems with allowance auctions)—can fully meet these criteria (in conjunction with complementary measures, such as research and development and investment in transportation infrastructure).

Fiscal instruments targeted directly at the sources of environmental harm promote the entire range of possibilities for reducing that harm. They can also produce a substantial revenue gain and, so long as this revenue is used productively—for example, to reduce other taxes that distort economic activity—environmental protection is achieved at lowest overall cost to the economy. Finally, and not least, if these instruments are scaled to reflect environmental damage, they avoid either excessively burdening the economy or, conversely, forgoing socially worthwhile environmental improvements.

Getting Prices Right

"Getting prices right" is convenient shorthand for the idea of using fiscal instruments to ensure that the prices that firms and consumers pay for fuel reflect the full costs to society of their use, which requires adjusting market prices by an appropriate set of "corrective" taxes. In practice, many countries, far from charging for environmental damage, actually subsidize the use of fossil fuels. For many others, energy taxes—if currently implemented at all—are often not well targeted at sources of environmental harm, nor set at levels that appropriately reflect environmental damage. Clearly there is much scope for policy reform in this area, but there are also huge challenges, both practical and analytical.

From a practical perspective, higher energy prices burden households and firms and, even with well-intentioned compensation schemes, can be fiercely resisted. These challenges—not to understate them—are largely beyond the scope of this book; however, a complementary volume (Clements and others, 2013) distills lessons to be drawn from case studies of energy price reforms. Moreover, getting energy prices right need not increase the overall tax burden; higher fuel taxes could partially replace broader taxes on income or consumption (or environmentally blunt taxes on energy), broadening support for the policy. Where new revenue sources might be needed, corrective energy taxes are an especially attractive option because, unlike most other options, they improve economic efficiency by addressing a market failure.

The main focus here is on assessing the analytical challenges, that is, the pricing that needs to be put into practice. For the vast majority of countries, there has been no attempt to measure the magnitude of environmental damage across fossil fuel products—yet these measures are critical for actionable guidance to be given on how countries can get energy prices right.

The corrective energy tax estimates presented in this book should be treated with a good deal of caution, given data gaps, and controversies—for example, about the valuation of climate damage and the link between air quality and mortality risk. Nonetheless, the estimates provide a valuable starting point for dialogue about policy reform, scrutiny of the key uncertainties, and cross-country comparisons estimated on a consistent basis.[1] Moreover, the impact of alternative assumptions on corrective tax estimates can be derived from accompanying spreadsheets.[2] Although tax assessments may change significantly as evidence evolves and data improve, the basic findings—most notably, the strong case for substantially higher taxes on coal and motor fuels in many countries—are likely to remain robust.

Defining an Efficient Set of Energy Taxes

From the perspective of effectively reducing energy-related CO_2 emissions, local air pollution, and broader side effects from vehicle use, energy tax systems should comprise three basic components:

- A charge should be levied on fossil fuels in proportion to their CO_2 emissions multiplied by the global damage from those emissions (alternatively, the charge could be levied directly on emissions), though there are reasons why some governments (e.g., in low-income, low-emitting countries) may not wish to impose such charges.
- Additional charges should be levied on fuels used in power generation, heating, and by other stationary sources in proportion to the local air pollution emissions from these fuels but with credits for demonstrated emissions capture during fuel combustion, given that net emissions released are what determine environmental damage (another possibility again is to charge emissions directly).
- Additional charges for local air pollution, congestion, accidents, and pavement damage attributable to motor vehicles. Ideally, some of these charges would be levied according to distance driven (e.g., at peak period on busy roads for congestion), and doing so should become increasingly feasible as

---

[1] The estimates take into account extensive feedback from leading specialists and from participants at an expert workshop held at the IMF in April 2013.
[2] See www.imf.org/environment.

the technology needed for such programs matures. Until then, however, reflecting all of these costs in motor fuel taxes is appropriate and is the approach taken here.

In practice, there are complex political reasons why the bases and rates of energy taxes may diverge from the ideal, and why regulatory instruments are often the preferred approach. But a necessary first step for understanding the trade-offs involved between all the policy choices, and how political constraints might be met with minimal compromise to environmental, fiscal, or other objectives, is to provide some quantitative sense of the corrective energy tax system for different countries, which provides a benchmark against which alternative policies should be evaluated.

METHODOLOGY

The techniques used for assessing various types of environmental damage are straightforward conceptually, although they require extensive data compilation.

Climate Damage

The volume does not add to the contentious debate on climate damage but simply uses an illustrative damage value of $35 per metric ton of CO_2 (US Interagency Working Group on the Social Cost of Carbon, 2013), combined with data on the carbon content of fuels, to derive carbon charges for all countries. (The implications of different damage assumptions, including zero damage for low-income, low-emitting countries, are easily inferred.)

Air Pollution

The major problem from local air pollution is elevated mortality risks for exposed populations. For coal plant emissions, damage is assessed by first estimating how much pollution is inhaled by people in different countries based on combining data on power plant location with data indicating how many people live at different distance classifications from each plant (smokestack emissions can be transported over considerable distances). This pollution intake is then combined with baseline mortality rates for pollution-related illness and the latest evidence on the relationship between exposure and elevated risks, though substantial uncertainties surround this relationship. Health effects must then be monetized, which is a contentious exercise, but is done for illustrative purposes using evidence on how people in different countries value the trade-off between money and risk from numerous studies analyzed in OECD (2012). Finally, damage is expressed per unit of energy content or fuel use using country-level data on emission rates. The same approach is used to measure air pollution damage from natural gas plants. Damage from vehicle and other ground-level sources (which tend to remain locally concentrated) is extrapolated from a city-level database on pollution intake rates.

Congestion and Accidents

Traffic congestion costs imposed by one driver on other vehicle occupants are approximated by using a city-level database to estimate relationships between travel delays and various transportation indicators and extrapolating the results using country-level measures of those same indicators. Travel delays are monetized using evidence about the relationship between wages and how people value travel time. Accident costs are estimated based on country-level fatality data and assumptions about which types of risks drivers themselves might take into account versus those they do not, and extrapolations of various other costs, such as those for medical expenses, property damage, and nonfatal injury.

MAIN FINDINGS

The main policy messages include the following:

- **Coal use is pervasively undercharged, not only for carbon emissions, but also for the health costs of local air pollution.** Illustrative charges for CO_2 amount to about $3.3/gigajoule (GJ) of energy from coal combustion, a substantial sum when set against average world coal prices of about $5/GJ in 2010, and what are at best minimal taxes on coal at present.[3] The corrective taxes for local air pollution can also be substantial—they exceed the carbon charge for 10 of the 18 countries illustrated in panel 1 of Figure 1.1—though they vary considerably across countries with population exposure, emission rates, and the value of mortality risks. These corrective taxes for local air pollution are, however, based on current emission rates averaged across plants with and without control technologies. Appropriate crediting would provide strong incentives for all plants to adopt control technologies.

- **Air pollution damage from natural gas is modest relative to that from coal, but significant tax increases are still needed to reflect carbon emissions.** Corrective charges for local air pollution from natural gas are about $1/GJ or less for most countries shown in panel 2 of Figure 1.1. The carbon component is also smaller: natural gas produces about 40 percent less CO_2 per GJ than coal, though charges needed to cover carbon emissions, about $2/GJ (40 percent of average world gas prices), are well above current tax levels.

- **Higher taxes on motor fuels are warranted in many countries, though more to reflect the costs of traffic congestion and accidents than carbon emissions and local air pollution.** Corrective gasoline taxes are about $0.40/liter (about $1.50/gallon) or more in 17 of the 20 countries illustrated

---

[3] All corrective tax estimates are for 2010 and, to facilitate cross-country comparisons, are expressed in 2010 U.S. dollars. Purchasing-power-parity (PPP) exchange rates should be used to express these figures in local currency (see www.imf.org/external/pubs/ft/weo/2013/01/weodata/index.aspx).

Figure 1.1 Corrective Fuel Taxes to Reflect Environmental Costs, Selected Countries, 2010

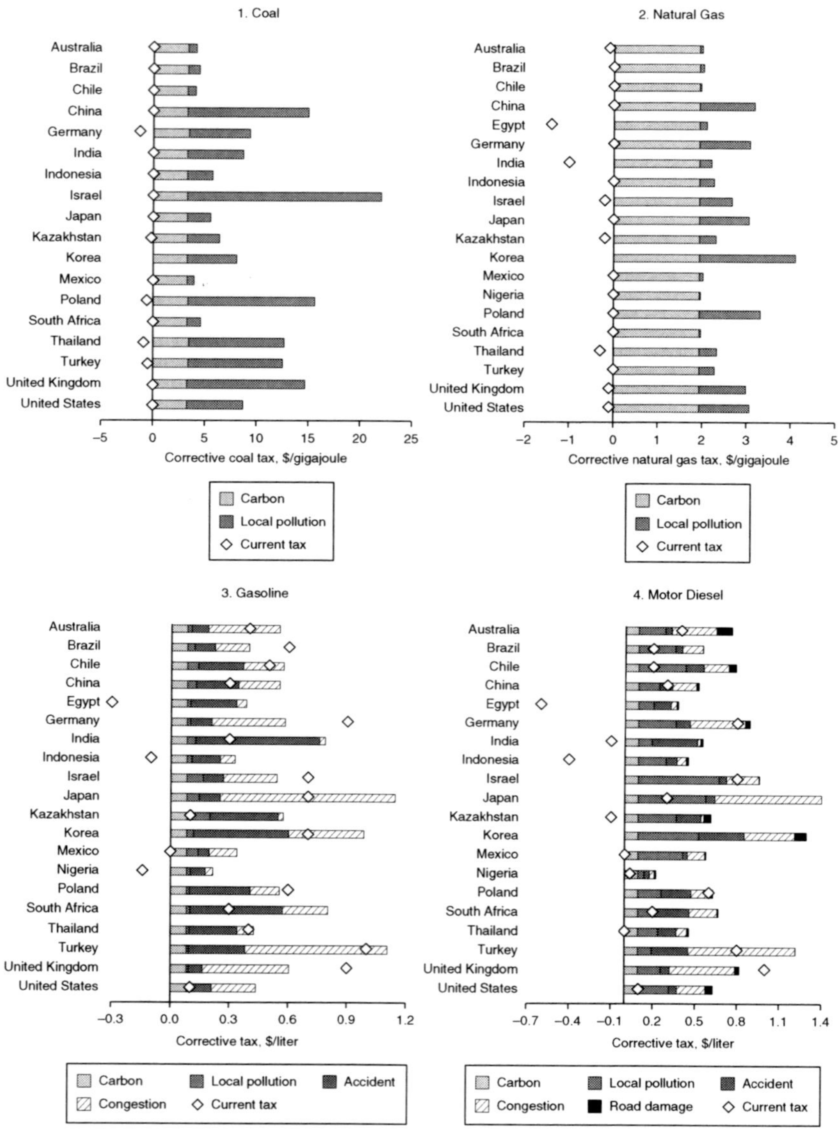

Source: Authors, based on methodology described in the book.

in panel 3 of Figure 1.1, and they exceed current excise taxes in 15 of these countries, with congestion and accidents costs together accounting for about 70–90 percent of the corrective tax. CO_2 emissions contribute $0.08/liter in all cases and local pollution typically less. Corrective taxes for diesel fuel (primarily used by trucks but also by some cars) are often somewhat higher than for gasoline (panel 4 of Figure 1.1), so there appears to be little

Figure 1.2 Impacts of Fuel Tax Reform, Selected Countries, 2010

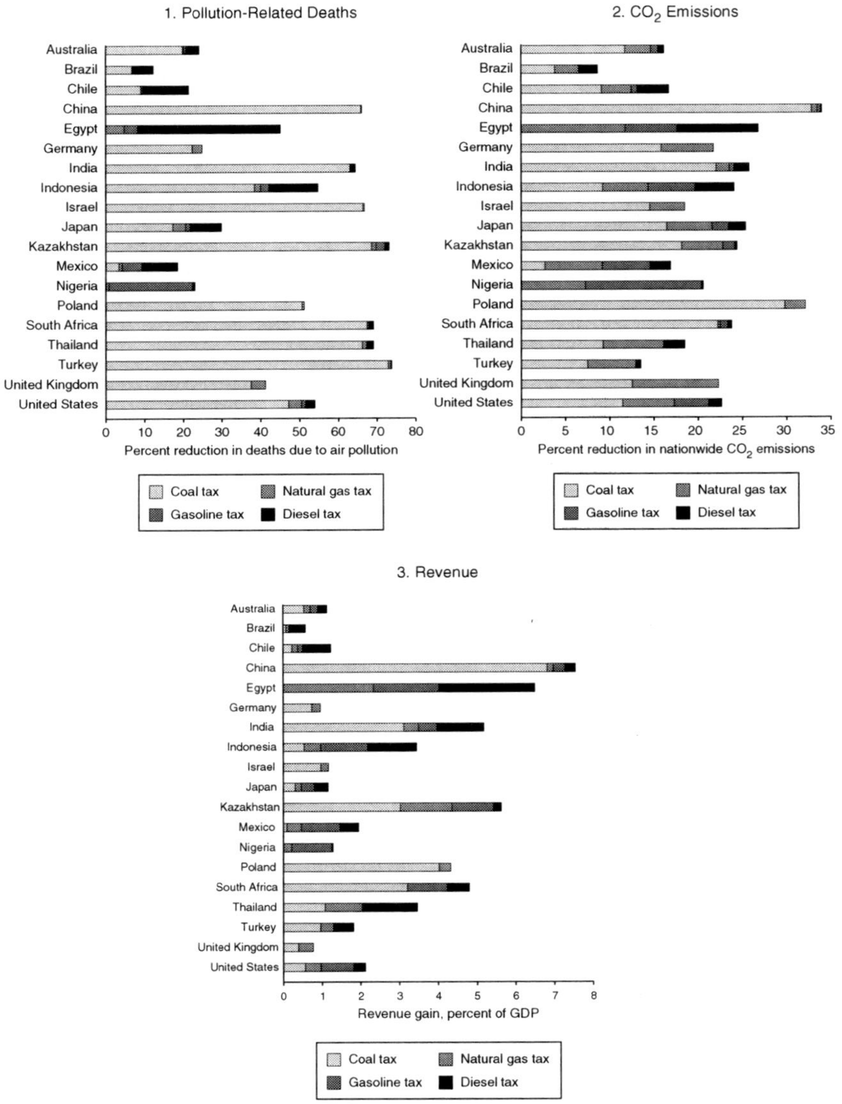

1. Pollution-Related Deaths

2. CO_2 Emissions

3. Revenue

Source: Authors, based on methodology described in the book.

systematic basis for the common practice of taxing motor diesel at lower rates than gasoline.

- **Corrective taxes can yield substantial reductions in pollution-related deaths and in CO_2 emissions, and large revenue gains**:
  - *Fuel tax reform can reduce worldwide deaths from outdoor, fossil fuel, air pollution by 63 percent.* The vast majority of avoided deaths for countries shown in panel 1 of Figure 1.2 result from corrective taxes on coal.

- ***Tax reforms could reduce CO_2 emissions by 23 percent globally.*** For all but five countries shown in panel 2 of Figure 1.2, coal (because of its high carbon intensity and high corrective taxes) accounts for more than 50 percent of the CO_2 reductions, and about 75 percent or more in seven countries.
- ***Potential revenue from implementing corrective taxes averages 2.6 percent of GDP globally.*** Corrective taxes on coal could be a significant revenue source for many countries shown in panel 3 of Figure 1.2, especially coal-intensive ones such as China (though revenue projections are necessarily very approximate). In other countries, such as Brazil, Egypt, Indonesia, Japan, Mexico, Nigeria, and the United States, higher motor fuel taxes are the dominant source of potential revenue gains (including subsidy elimination in some cases).
- ***In short, the case for substantially higher energy taxes does not rest on climate change alone.*** Decisive action need not wait on global coordination.

SUMMARY AND CONCLUSION

Getting energy prices right involves a straightforward extension of widely accepted and easily administered motor fuel taxes—better aligning the rates of these taxes with environmental damage and extending similar charges to other fossil fuel products (or their emissions). There are complications (e.g., charges should be based on emissions net of any application of emissions control technologies), but the issues should be manageable. The findings of this volume suggest large and pervasive disparities between efficient fuel taxes and current practice in developed and developing countries alike, with much (in fact, a huge amount in many countries) at stake for fiscal, environmental, and health outcomes.

The main challenge is how to get it done—how to build support for energy price reform. International organizations and others have an important role to play, first in promoting dialogue about best practice, and second in providing solid analytical contributions quantifying the benefits of pricing policies relative to alternative approaches, and assessing distributional implications to inform the design of compensating measures.

OUTLINE OF THE VOLUME

The main findings of the report—the corrective tax estimates by fuel and by country, and rough estimates of the fiscal, environmental, and health benefits from tax reform—are presented in Chapter 6, which can be read without reading the preceding chapters for those who prefer to go straight to these results. The other chapters are organized as follows: A quick overview of energy systems, the nature of environmental side effects, and major fiscal policies affecting energy is provided in Chapter 2. Chapter 3 describes the case for and design of fiscal instruments to address environmental side effects. Chapter 4 then discusses the measurement of global, and in particular local, air pollution damage from fossil fuel use. The measurement of congestion, accident, and road damage costs associated with vehicle use is discussed in Chapter 5. Chapter 7 offers brief concluding remarks.

REFERENCES

Clements, Benedict, David Coady, Stefania Fabrizio, Sanjeev Gupta, Trevor Alleyene, and Carlo Sdralevich, eds., 2013, *Energy Subsidy Reform: Lessons and Implications* (Washington: International Monetary Fund).

Intergovernmental Panel on Climate Change (IPCC), 2013, *Climate Change 2013: The Physical Science Basis*, Contribution of Working Group I to the Fifth Assessment Report of the Intergovernmental Panel on Climate Change (Cambridge, U.K.: Cambridge University Press).

National Research Council, 2009, *Hidden Costs of Energy: Unpriced Consequences of Energy Production and Use* (Washington: National Research Council, National Academies).

Organization for Economic Cooperation and Development, 2012, *Mortality Risk Valuation in Environment, Health and Transport Policies* (Paris: Organization for Economic Cooperation and Development).

Parry, Ian W.H., and Kenneth A. Small, 2009, "Should Urban Transit Subsidies Be Reduced?" *American Economic Review*, Vol. 99, No. 3, pp. 700–24.

World Bank and State Environmental Protection Agency of China, 2007, *Cost of Pollution in China: Economic Estimates of Physical Damages* (Washington: World Bank).

World Health Organization (WHO), 2013, *Global Health Observatory Data Repository* (Geneva: World Health Organization).

CHAPTER 2

Energy Systems, Environmental Problems, and Current Fiscal Policy: A Quick Look

Fossil fuels are used pervasively to generate electricity, power transportation vehicles, and provide heat for buildings and manufacturing processes. Fuel combustion produces carbon dioxide (CO_2) emissions and various local air pollutants, and use of transportation vehicles also causes road congestion, accidents, and (less important) pavement damage.

This chapter provides a quick look at energy systems, elaborates on their major environmental impacts, and discusses existing fiscal provisions affecting energy. Although the information here is not directly relevant for estimating corrective fuel taxes, it provides broader context and suggests why corrective taxes, and their impacts, are likely to differ considerably across countries.

OVERVIEW OF ENERGY SYSTEMS

Although insofar as possible this volume presents results for 156 countries, a focus on 20 countries is used to illustrate how corrective taxes and their impacts vary with per capita income, fuel mixes, population density, road fatalities, and so on. This section provides some basic statistics for these countries for 2010 (or the latest year for which data are available).

Figure 2.1 shows primary energy consumption (i.e., the energy content of fossil and other fuels before transformation into electricity) in gigajoules per capita. Energy consumption is highest in the United States and roughly half as high in countries such as Germany, Japan, and the United Kingdom. At the other end of the spectrum, energy consumption per capita in India, Indonesia, and Nigeria is 8 percent or less of that in the United States.

These differences primarily reflect variations in reliance on electricity and motor vehicles. As indicated in Figure 2.2, relative differences in electricity consumption per capita broadly follow the patterns for total energy consumption per capita. In the United States, for example, people tend to live in relatively large homes requiring higher electricity use, whereas in Indonesia and India, about 35 percent of the population lacks access to electricity, as do about 50 percent in Nigeria (World Bank, 2013).

Similarly, countries with lower per capita energy consumption also tend to have lower vehicle ownership rates (Figure 2.3). The United States and Australia, for

12 Getting Energy Prices Right: From Principle to Practice

Figure 2.1 Primary Energy Consumption per Capita, Selected Countries, 2010

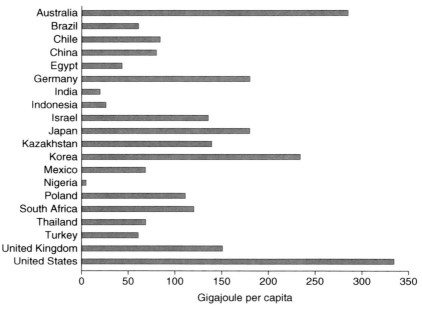

Source: US EIA (2013).
Note: Primary energy consumption is the energy content of fossil and other fuels before transformation into power generation.

Figure 2.2 Electricity Consumption per Capita, Selected Countries, 2010

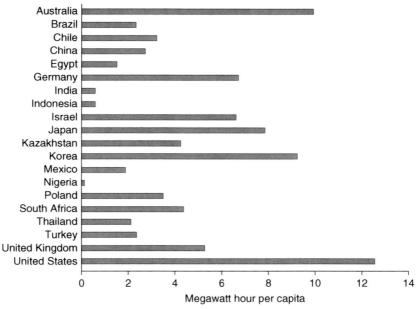

Source: US EIA (2013).
Note: Electricity consumption includes residential and industrial uses.

Figure 2.3 Motor Vehicle Ownership Rates, Selected Countries, 2010 (or latest available)

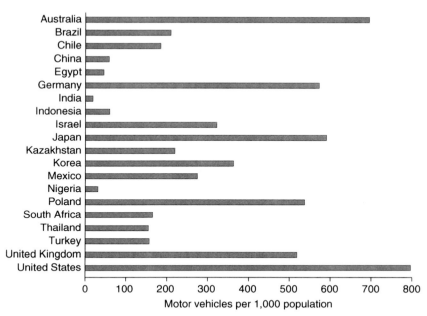

Source: World Bank (2013).
Note: Motor vehicles include cars, trucks, and buses. However, two-wheeled motorized vehicles (which are used pervasively in many Asian countries) are not included in the data.

example, have about 800 and 700 motor vehicles per thousand people, respectively, whereas China, Egypt, India, Indonesia, and Nigeria, have fewer (sometimes far fewer) than 100 vehicles per thousand people.

The scale of environmental problems also depends critically on a country's fuel mix, and again there are large differences, as indicated in Figure 2.4. For example, coal constitutes more than half of total energy consumption in China, India, Kazakhstan, Poland, and South Africa, but 5 percent or less in Brazil, Egypt, Mexico, and Nigeria. Petroleum varies from 19 percent of energy consumption in China to 71 percent in Nigeria. And natural gas varies from 2 percent in South Africa to 50 percent in Egypt. Countries use some renewables (wind, solar, hydro, and others), but there are challenges to their growth, such as the intermittent supply from wind and solar power and the mismatch between their ideal location and urban centers.

Differences in energy consumption per capita, in particular, but also in fuel mixes, explain differences in energy-related CO_2 emissions per capita, shown in Figure 2.5. For example, annual emissions per capita are almost 20 metric tons in Australia and the United States, countries that both use a lot of energy and have relatively emissions-intensive fuel mixes.

The severity of environmental problems also depends on population density (greater density generally indicates that more people are exposed to local air emissions and that road systems are more crowded), which again varies considerably

Figure 2.4 Share of Final Energy Use by Fuel Type, Selected Countries, 2010 (or latest available)

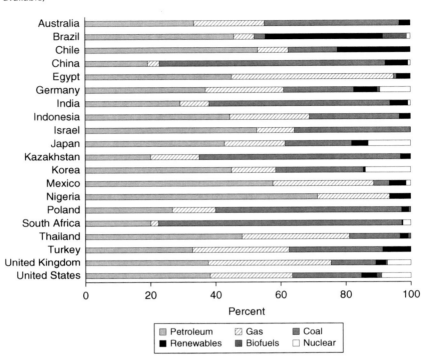

Source: US EIA (2013).
Note: The figure shows the share of primary energy (from direct fuel combustion) and secondary energy (primarily power generation) attributed to different fuels. Fuels are compared on an energy-equivalent basis.

across countries. For example, the share of the population living in urban areas ranges from about 90 percent in Australia, Chile, Israel, and Japan to less than 40 percent in India and Thailand (Figure 2.6).

ENVIRONMENTAL SIDE EFFECTS

Fossil fuel use is associated with a variety of environmental side effects, or "externalities." An adverse externality occurs when the actions (e.g., fuel combustion) of individuals or firms impose costs on others that the actors do not take into account. Externalities call for policy intervention, principally monetary charges that are directly targeted at the source of environmental harm and that are set at levels to reflect environmental damage (Chapter 3).

The main externalities of concern for this study are CO_2 emissions, local air pollution, and the broader costs of vehicle use. Further environmental problems are discussed in Box 2.1.

Figure 2.5 Carbon Dioxide (CO_2) Emissions per Capita, Selected Countries, 2010

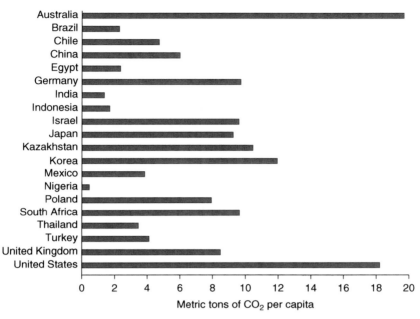

Source: US EIA (2013).

Figure 2.6 Urban Population, Selected Countries, 2010

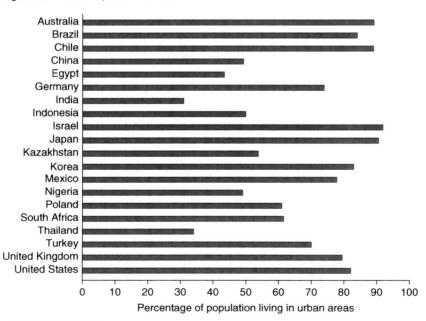

Source: World Bank (2013).
Note: Urban population refers to the share of people living in urban areas as defined by national statistical offices.

BOX 2.1

Broader Environmental Effects beyond the Study Scope

A variety of other costs associated with fossil fuel production and use are not considered in this study for one or more of the following reasons (see, e.g., NRC, 2009, Chapter 2, for further discussion):

- These costs might be taken into account by individuals and firms
- They may be modest in quantitative terms
- They might be too difficult to quantify
- They may call for policies other than fuel taxes

Examples include the following:

Additional pollutants: Carbon monoxide (CO), a by-product of fuel combustion, reduces oxygen in the bloodstream, posing a danger to those with heart disease, but when released outdoors its concentration is usually not sufficient to cause significant health effects. Lead emissions cause neurological effects, especially for children, with potentially significant impacts on lifetime productivity (Grosse and others, 2002; Zax and Rees, 2002). However, lead has been, or is being, phased out from petroleum products in many countries. Various other toxins (e.g., benzene) are generally not released in sufficient quantities to cause health damage that is significant relative to the effects of pollutants considered in this book.

Upstream environmental impacts: Environmental impacts occurring during fuel extraction and production include

- Despoiling of the natural environment (e.g., mountaintop removal for coal, accidents at oil wells)
- Waste from fuel processing (e.g., slurry caused by the "washing" of raw coal)
- Emissions leaks during fuel storage (e.g., from corrosion or evaporation at underground tanks at refineries and gasoline stations)
- Further leakage during transportation (e.g., spills from oil tankers)

However, per unit of fuel use, the damage from these causes appear to be small relative to those estimated in this study (Jaramillo, Griffin, and Matthews, 2007; NRC, 2009, Chapter 2) and these problems call for interventions (e.g., double hull requirements for tankers, mandatory insurance for accident costs, requirements that mined areas be returned to their premining vegetative state) other than fuel taxes.

Occupational hazards: For fossil fuel extraction industries, occupational hazards include, for example, lung disease from long-term exposure to coal dust, coal mine collapses, and explosions at oil rigs. Individuals may account for these risks, however, when choosing among different occupations (a long-established literature in economics suggests that higher-risk jobs tend to compensate workers through higher wages; see, e.g., Rosen, 1986). To the extent policy intervention is warranted, perhaps because individuals understate risks, more targeted measures such as workplace health and safety regulations would be more efficient than fuel charges.

Indoor air pollution: Indoor air pollution causes an estimated 3.8 million deaths worldwide each year (Burnett and others, 2013). For example, in low-income countries, burning coal in poorly ventilated cooking stoves or open fires can create serious pollution-related health problems (Ezzati, 2005). Raising consumer coal prices may not be the best policy for dealing with indoor air pollution, however, particularly because doing so may promote the equally harmful use of biomass, at least until cleaner energy sources (e.g., charcoal, natural gas, electricity, or even processed coal that burns more cleanly), and better technologies such as better ventilated stoves, are made available.

Energy security: Although energy security concerns often motivate policies to reduce domestic consumption of oil and other fuels, quantifying a reasonable fuel tax level for this purpose is challenging. Some studies (e.g., Brown and Huntington, 2010) suggest the costs (not taken into account by the private sector) arising from the vulnerability of the macroeconomy to oil price volatility are not especially large, at least for the United States. More generally, dependence on oil supplies from politically volatile regions may realign a country's foreign policy away from globally desirable objectives toward one focused on promoting access to oil markets (Council on Foreign Relations, 2006), though rapid development of unconventional oil, such as shale oil, may be alleviating these concerns.

CO_2 Emissions

CO_2 emissions from fossil fuel combustion are, by far, the largest source of global greenhouse gas (GHG) emissions. Emissions trends and the scientific basis for human-induced global warming are briefly summarized in this section. For an in-depth discussion, see successive reports of the Intergovernmental Panel on Climate Change (IPCC), most recently the assessment of the science in IPCC (2013).

Global, energy-related CO_2 emissions have increased from about 2 billion metric tons in 1900 to about 30 billion tons in 2013 and, in the absence of mitigating measures, are projected to increase to almost 45 billion tons by 2035 (Figure 2.7). Emissions from non-OECD countries overtook those from OECD countries about 2005, and are projected to account for two-thirds of the global total by 2035.

Roughly 50 percent of CO_2 releases accumulate in the global atmosphere, where they remain, on average, for about 100 years; consequently, atmospheric CO_2 concentrations have increased from pre-industrial levels of about 280 parts per million (ppm) to current levels of about 400 ppm. Accounting for other GHGs, such as methane and nitrous oxide (from agricultural and industrial sources), and expressing them in lifetime warming equivalents to CO_2, atmospheric concentrations in CO_2 equivalents are now about 440 ppm. In the absence of substantial emissions mitigation measures, GHG concentrations are expected to reach 550 ppm (in CO_2 equivalent) by about the middle of this century, and to continue rising thereafter (Aldy and others, 2010; Bosetti and others, 2012).

IPCC (2013) estimates that global average temperatures have risen by 0.85°C since 1880 and is 95 percent certain that the main cause is fossil fuel

Figure 2.7 Projected Global Energy-Related CO_2 Emissions

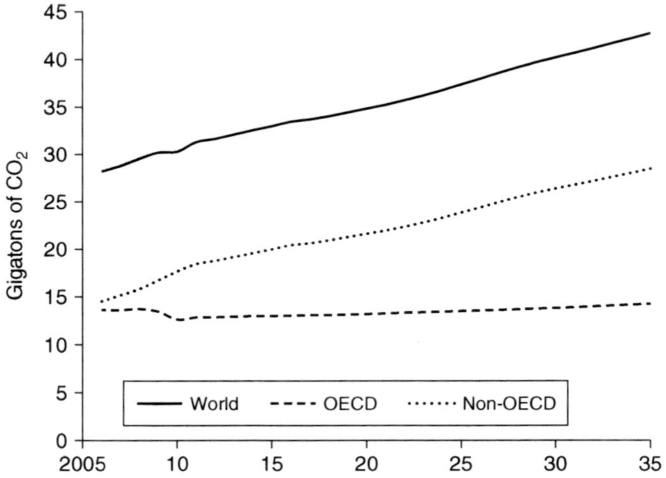

Source: US EIA (2011, Table A10).
Note: CO_2 = carbon dioxide; OECD = Organization for Economic Cooperation and Development.

combustion and other man-made GHGs (rather than other factors like changes in solar radiation and heat absorption in urban areas). However, because of lags in the climate system (i.e., gradual heat diffusion processes in the oceans), temperatures are expected to continue rising, even if concentrations were to be stabilized at current levels. As indicated in Figure 2.8, if GHG concentrations were stabilized at 450, 550, or 650 ppm, respectively, the eventual mean projected warming (over pre-industrial levels) is 2.1, 2.9, and 3.6°C, respectively.[1] Alternatively, contemporaneous warming is expected to reach about 3–4°C by the end of the century, though actual warming could be substantially higher (or lower) than this (Bosetti and others, 2012; IPCC, 2013; Nordhaus, 2013, Figure 9).

The climatic consequences of warming include changed precipitation patterns, sea level rise caused by thermal expansion of the oceans and melting sea ice, more intense and perhaps frequent extreme weather events, and possibly more catastrophic outcomes like runaway warming, ice sheet collapses, or destruction of the marine food chain caused by warmer, more acidic oceans. Considerable uncertainty surrounds all of these effects, particularly from the potential for feedback effects (e.g., releases of methane from thawing permafrost tundra, less reflection of sunlight as glaciers melt) that might compound warming.

---

[1] For perspective on the scale of these changes, current temperatures are about 5°C higher than at the peak of the last ice age about 20,000 years ago when the climate was radically different and much of the northern hemisphere was covered in ice.

Figure 2.8 Projected Long-Term Warming above Pre-Industrial Temperatures from Stabilization at Different Greenhouse Gas Concentrations

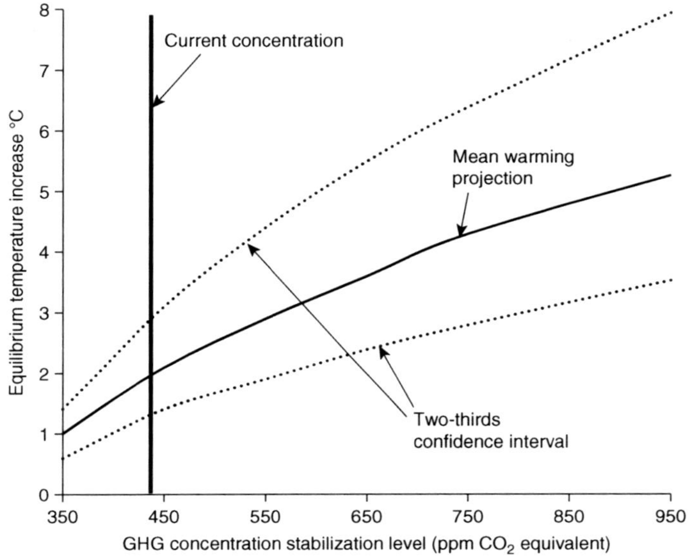

```
[Chart: Equilibrium temperature increase °C (y-axis, 0–8) vs. 
GHG concentration stabilization level (ppm CO₂ equivalent) (x-axis, 350–950).
Shows: Current concentration (vertical line), Mean warming projection (solid curve), 
Two-thirds confidence interval (dotted curves).]
```

Source: IPCC (2007, Table 10.8).

Note: CO_2 = carbon dioxide; GHG = greenhouse gas; ppm = parts per million. Figure shows the projected increase in global temperature (once the climate system has fully adjusted, which takes at least several decades) over pre-industrial levels if atmospheric GHG concentrations are stabilized at different levels. The most recent assessment (IPCC, 2013) slightly lowered the bottom end of the confidence interval for a doubling of CO_2 equivalent (not reflected in the figure).

Policies to price the carbon content of fossil fuels (or otherwise mitigate CO_2 emissions) are needed, because at present households and firms are generally not charged for the future climate change damage resulting from these emissions.[2]

Local Air Pollution

Unless it is priced to reflect environmental damage, local air pollution from fuel combustion is also excessive from society's perspective. The sources of air pollution and their environmental impacts are discussed in this section.

Sources of air pollution

Fossil fuel use results in both primary pollutants, emitted during fuel combustion, and secondary pollutants, formed subsequently from chemical transformations of

---

[2] Other policies are also needed, but are largely beyond the scope of this book. These include policies to reduce emissions from international aviation and maritime activities (Keen, Parry, and Strand, 2013) and land use (Mendelsohn, Sedjo, and Sohngen, 2012); measures to enhance clean technology development (see Chapter 3); adaptation to climate change (e.g., coastal defenses, shifting to hardier crop varieties); development of last-resort technologies (e.g., to remove CO_2 from the atmosphere or to "manage" solar radiation through sunlight-deflecting particles) for possible deployment in extreme scenarios; and mobilization of financial assistance for developing countries (de Mooij and Keen, 2012).

primary pollutants in the atmosphere. With regard to pollution-related health effects—the main manifestation of environmental damage—potentially the most important pollutant is fine particulates or $PM_{2.5}$ (particulate matter with diameter up to 2.5 micrometers) because these particles permeate the lungs and bloodstream. Primary $PM_{2.5}$ is directly emitted when fuels like coal are combusted, but is also formed indirectly from chemical reactions in the atmosphere involving certain primary pollutants.

The most important pollutants associated with coal are directly emitted $PM_{2.5}$ and sulfur dioxide (SO_2), which reacts in the atmosphere to form $PM_{2.5}$. Fine particulates are also formed from nitrogen oxide (NO_x) emissions, but generally in smaller quantities, because NO_x emission rates are generally lower than for SO_2 and are less reactive. Emission rates per unit of energy can vary considerably across different coal types, and a number of newer coal plants in many countries incorporate emissions control technologies. (Both factors should be considered in setting coal taxes.)

Natural gas is a much cleaner fuel than coal; it produces a minimal amount of SO_2 and primary $PM_{2.5}$ emissions, though it does generate significant amounts of NO_x. Motor fuel combustion also produces NO_x, and diesel fuel combustion causes some SO_2 and primary $PM_{2.5}$ emissions. Motor fuel combustion also releases volatile organic compounds that react with NO_x in the presence of sunlight to form ozone, a major component of urban smog. Ozone has health effects, though the link to mortality is much weaker than for $PM_{2.5}$ (damage from ozone is not considered here).[3]

Environmental damage

Local air pollution damage is potentially large, and has been estimated to be about 1 percent of GDP for the United States and almost 4 percent for China (NRC, 2009; Muller and Mendelsohn, 2012; World Bank and State Environmental Protection Agency of China, 2007). These harmful effects range from impaired visibility and nonfatal heart and respiratory illness to building corrosion and reduced agricultural yields when pollutants react with water to form acid rain. However, a number of studies suggest that, by far, the main damage component (and the component this study focuses on) is elevated risks of premature human mortality.[4]

The epidemiological literature has solidly established that long-term exposure to $PM_{2.5}$ is associated with increased risk of lung cancer, chronic obstructive pulmonary disease, heart disease (from reduced blood supply), and stroke (Burnett and others, 2013; Health Effects Institute, 2013; Humbert and others, 2011; Krewski and others, 2009). Seniors, infants, and people with preexisting

---

[3] This ground-level ozone is distinct from stratospheric ozone, which blocks cancer-causing, ultraviolet radiation. Stratospheric ozone depletion is caused by man-made chemicals, but such chemicals have now been largely phased out (Hammitt, 2010).

[4] For example, studies for China, Europe, and the United States find that mortality impacts typically account for 85 percent or more of the total damage from local air pollution (US EPA, 2011; European Commission, 1999; NRC, 2009; World Bank and State Environmental Protection Agency of China, 2007; Watkiss, Pye, and Holland, 2005).

Figure 2.9 Air Pollution Concentrations, Selected Countries, 2010

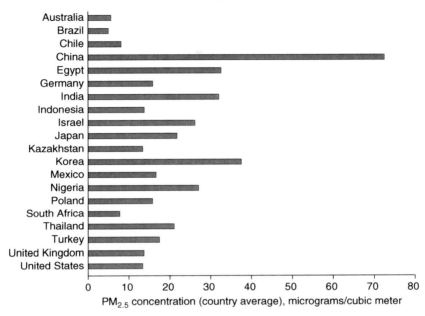

Source: Brauer and others (2012).
Note: PM$_{2.5}$ = fine particulate matter. Data are averages of regional pollution concentrations (weighted by population shares) within a country; regional observations are based on satellite data. Concentrations for specific urban centers can be much higher than national averages.

health conditions (e.g., those who have suffered strokes or who are suffering from cardiovascular disease) are most susceptible (Rowlatt and others, 1998).

Figure 2.9 shows ambient PM$_{2.5}$ concentrations in 2010 for selected countries (averaged across regional pollution concentrations in each country after weighting regions by population shares). For many countries (e.g., Germany, Indonesia, Kazakhstan, Mexico, Poland, the United Kingdom, and the United States) average PM$_{2.5}$ concentrations are between about 10 and 20 micrograms/cubic meter. Some countries have average PM$_{2.5}$ concentrations of less than 10 micrograms/cubic meter (e.g., Australia, Brazil, and South Africa, where the coastal location of cities helps to disperse pollution). But in other countries, PM$_{2.5}$ concentrations can be much greater; for example, between 30 and 40 micrograms/cubic meter in Egypt, India, and Korea and, strikingly, more than 70 micrograms/cubic meter in China.

Figure 2.10 shows estimated deaths by region attributable to local outdoor air pollution in 2010. Worldwide, deaths were 3.2 million and were especially concentrated in East Asia (about 1.3 million) and South Asia (about 0.8 million).

Fuel taxes should not necessarily be highest in countries with the worst pollution, however, because the extra health risks posed by additional pollution do not necessarily depend on existing pollution concentrations (Chapter 3). Appropriate taxes depend, for example, on the size and composition of the exposed population and on

Figure 2.10 Air Pollution Deaths by Region, 2010

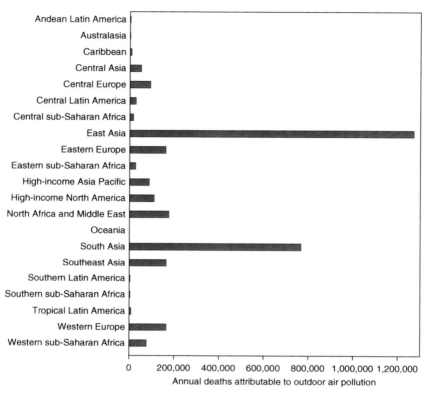

Source: Burnett and others (2013).
Note: Figure shows estimated deaths from outdoor ambient air pollution and excludes deaths from indoor air pollution (see Box 2.1). Data by country were not available at the time of writing.

how health risks are valued (which might vary with income levels). A given level of environmental tax, however, likely has relatively larger environmental impacts in high-pollution countries, where there is greater scope for reducing pollution.

Broader Externalities Related to Motor Fuel Use

Use of fuel in motor vehicles is associated with further side effects that should be factored into tax design, the most important of which are traffic congestion and traffic accidents. (Road damage plays a more minor role in corrective fuel taxes.)

Traffic congestion

Traffic on roads where speeds are below free-flow levels is generally excessive: unless they are charged for road use, motorists will not account for their own impact on adding to congestion and slowing speeds for other road users (Arnott, Rave, and Schöb, 2005; Lindsey, 2006; Litman, 2013; Santos, 2004). This applies irrespective of complementary policies such as investment in road or transit capacity or improved coordination of traffic signals, though these improvements can lower the appropriate charge for congestion (e.g., by alleviating bottlenecks).

Traffic congestion varies dramatically across urban and rural areas, and across time of day. It has been estimated, for example, that drivers in the London rush hour impose costs on others equivalent to US$10 per liter of fuel through their contribution to traffic congestion (Parry and Small, 2009). Congestion is best addressed through taxes on vehicle-kilometers driven on busy roads, with rates varying over the course of the day with prevailing traffic levels (Chapter 3). Until such charges are comprehensively implemented (e.g., using global positioning systems), however, congestion costs that motorists impose on others should be reflected appropriately in fuel taxes (Parry and Small, 2005).

The appropriate fuel charges are likely to vary considerably across countries, even if travel delays were valued similarly. Figure 2.11, which shows registered vehicles (cars, trucks, buses) per kilometer of nationwide road capacity, provides some sense, albeit very crude, of these variations. For example, Germany, Japan, Mexico, Poland, and the United Kingdom have far more vehicles per kilometer of road capacity than the United States, implying that a much greater portion of nationwide driving likely occurs under congested conditions in those countries.

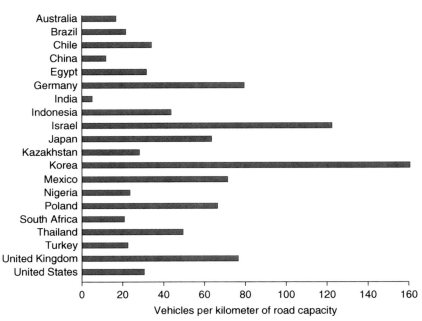

Figure 2.11 Vehicles and Road Capacity, Selected Countries, 2007 (or latest available)

Source: IRF (2009).
Note: Road capacity includes both paved and unpaved roads. Vehicles include cars, buses, and trucks but not motorized two wheelers.

Figure 2.12 Road Deaths, Selected Countries, 2010

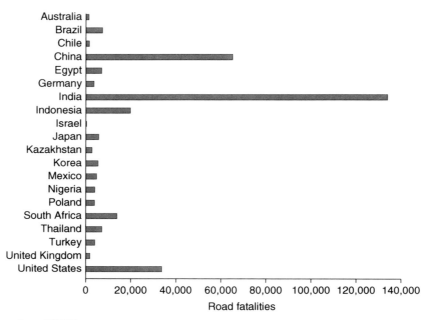

Source: IRF (2012).
Note: These data, which are for 2010 or the latest available year, may understate road fatalities in developing countries because of underreporting; see Chapter 5. WHO (2013) suggests, for example, that traffic deaths in India and China are much larger, and global total deaths are 1.2 million.

Traffic accidents

Another side effect of vehicle use is traffic accidents. Although drivers should take into account some accident costs, such as injury risks to themselves, other costs (e.g., injury risks to pedestrians, property damage, medical costs borne by third parties) are not taken into account, implying excessive driving from a societal perspective. Again, this applies irrespective of other measures, such as drunk driver penalties, airbag and seatbelt mandates, and traffic medians, though these measures lower the appropriate charge for accidents (e.g., by reducing fatality rates).

Figure 2.12, which shows the number of road fatalities in 2010, gives some sense of the problem. In India, for example, there were about 134,000 road deaths; in China about 65,000; and even in South Africa (which has 4 percent of the population of China), there were about 14,000 fatalities.

FISCAL POLICIES CURRENTLY AFFECTING ENERGY AND TRANSPORTATION

For data quality reasons, the discussion of tax policies in this subsection focuses on OECD countries.[5] Among these countries, revenue from environment-related

---

[5] Estimates of taxes and subsidies by fuel product for all countries are provided in Annex 6.2.

Figure 2.13 Revenue from Environment-Related Taxes as Percent of Total Revenue in OECD Countries, 2010 (or latest available)

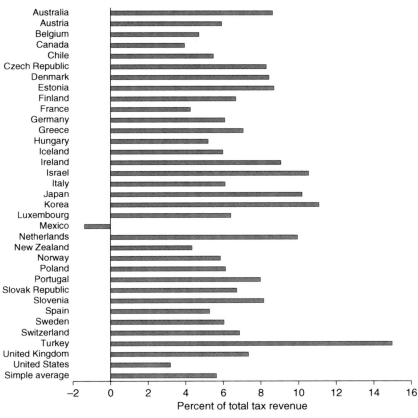

Source: OECD (2013).

taxes (Figure 2.13) averaged about 6 percent of total tax revenue in 2010, varying between 15 percent of revenue in Turkey, to 3 percent in the United States, and about minus 1.5 percent in Mexico, where petroleum was subsidized significantly in 2010 (before price liberalization in 2013).

These revenues mainly reflect three excise taxes: on fuel, on vehicle ownership, and on residential electricity consumption.

Although fuel taxes foster all possibilities for reducing fuel use (better fuel efficiency, less driving), the main issue in this analysis is whether tax levels reasonably reflect environmental damage. It seems unlikely that every country's taxes reflect that damage, given the huge disparities in tax rates (Figure 2.14). In 2010, gasoline taxes varied from the equivalent of more than US$0.80/liter in Finland, France, Germany, Greece, Israel, Norway, Turkey, and the United Kingdom, to US$0.11/liter in the United States, and a subsidy of US$0.13/liter in Mexico. In most countries, diesel fuel is tax-favored relative to gasoline, though it is not

Figure 2.14 Excise Tax Rates on Motor Fuels, 2010 (or latest available)

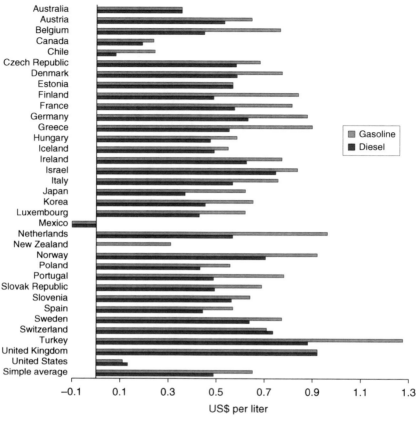

Source: OECD (2013).

obvious that trucks (the primary consumer of diesel) contribute less than cars to pollution, congestion, and so on.

Other taxes underlying Figure 2.13 are not well targeted from an environmental perspective (see Chapter 3). Vehicle taxes do not encourage vehicle owners to drive less and, despite often varying with emissions classes, do not exploit all opportunities for raising fuel efficiency. Simple taxes on electricity consumption do not encourage cleaner power generation fuels, nor use of emissions control technologies. Although carbon pricing is gathering momentum (Ecofys, 2013), as of 2013 about 80 percent of global CO_2 emissions were not covered by explicit carbon pricing programs, and CO_2 prices (currently equivalent to about US$7 per ton of CO_2 in the European Union Emissions Trading System) are typically a small fraction of estimated environmental damage (Chapter 5).

Moreover, many countries heavily subsidize—rather than tax—energy use. Estimated subsidies for fossil fuel use, measured by the gap between world fuel

Figure 2.15 Subsidies for Fossil Fuel Energy by Region and Fuel Type, 2011

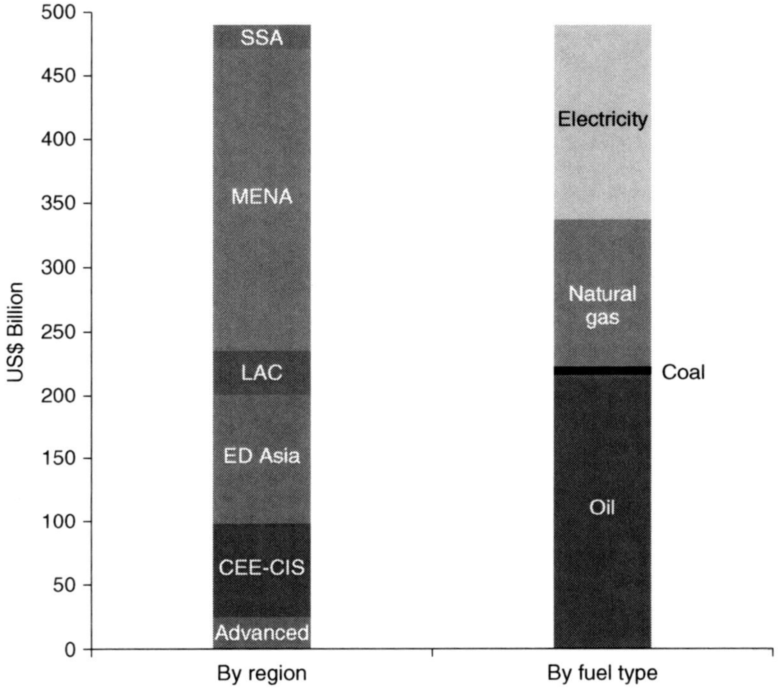

Source: Clements and others (2013).
Note: CEE-CIS = Central and Eastern Europe and the Commonwealth of Independent States; ED Asia = Emerging and Developing Asia; LAC = Latin America and the Caribbean; MENA = the Middle East and North Africa; SSA = sub-Saharan Africa.

prices and domestic market prices, were $490 billion worldwide in 2011, with the Middle East and North African countries accounting for 48 percent of these subsidies (Figure 2.15). Notably, 44 percent of these subsidies were for petroleum products, 23 percent for natural gas, 31 percent for electricity consumption, but only 1 percent for coal (the most polluting fuel)—so just eliminating subsidies without introducing coal taxes will have limited effects on emissions.[6]

Nonetheless, the overall picture is one of ample opportunity to rationalize energy prices by eliminating fossil fuel subsidies and shifting some of the burden of broader taxes onto fossil fuel products. Even countries with high energy taxes have scope to restructure them (e.g., by shifting taxes off electricity and onto coal) to improve their effectiveness, and to better align tax rates to environmental damage. How to gauge appropriate tax levels for this purpose is the main contribution of this volume.

---

[6]Renewables are also subsidized, at $66 billion in 2010, according to IEA (2011, Figure 14.13).

REFERENCES

Aldy, Joseph, Alan J. Krupnick, Richard G. Newell, Ian W.H. Parry, and William A. Pizer, 2010, "Designing Climate Mitigation Policy," *Journal of Economic Literature*, Vol. 48, pp. 903–34.

Arnott, Richard A., Tilmann Rave, and Ronnie Schöb, 2005, *Alleviating Urban Traffic Congestion* (Cambridge, Massachusetts: MIT Press).

Bosetti, Valentina, Sergey Paltsev, John Reilly, and Carlo Carraro, 2012, "Emissions Pricing to Stabilize Global Climate," in *Fiscal Policy to Mitigate Climate Change: A Guide for Policymakers*, edited by I.W.H. Parry, R. de Mooij, and M. Keen (Washington: International Monetary Fund).

Brauer, Michael, Markus Amann, Rick T. Burnett, Aaron Cohen, Frank Dentener, Majid Ezzati, Sarah B. Henderson, Michal Krzyzanowski, Randall V. Martin, Rita Van Dingenen, Aaron van Donkelaar, and George D. Thurston, 2012, "Exposure Assessment for Estimation of the Global Burden of Disease Attributable to Outdoor Air Pollution," *Environmental Science & Technology*, Vol. 46, pp. 652–60.

Brown, Stephen P.A., and Hillard G. Huntington, 2010, "Estimating US Oil Security Premiums," Discussion Paper 10–05 (Washington: Resources for the Future).

Burnett, Richard T., C. Arden Pope, Majid Ezzati, Casey Olives, Stephen S. Lim, Sumi Mehta, Hwashin H. Shin, and others, 2013, "An Integrated Risk Function for Estimating the Global Burden of Disease Attributable to Ambient Fine Particulate Matter Exposure" (Unpublished; Ottawa, Ontario, Canada: Health Canada).

Clements, Benedict, David Coady, Stefania Fabrizio, Sanjeev Gupta, Trevor Alleyene, and Carlo Sdralevich, eds., 2013, *Energy Subsidy Reform: Lessons and Implications* (Washington: International Monetary Fund).

Council on Foreign Relations, 2006, *National Security Consequences of US Oil Dependency* (Washington: Council on Foreign Relations).

de Mooij, Ruud, and Michael Keen, 2012, "Fiscal Instruments for Climate Finance," in *Fiscal Policy to Mitigate Climate Change: A Guide for Policymakers*, edited by I.W.H. Parry, R. de Mooij, and M. Keen (Washington: International Monetary Fund).

Ecofys, 2013, *Mapping Carbon Pricing Initiatives: Developments and Prospects.* Report produced for the World Bank (London: Ecofys).

European Commission, 1999, *ExternE Externalities of Energy, Vol. 7—Methodology Update*, Report produced for the European Commission, DG XII (Brussels: Office of Publications for the European Communities).

Ezzati, Majid, 2005, "Indoor Air Pollution and Health in Developing Countries," *Lancet*, Vol. 366, pp. 104–106.

Grosse, Scott D., Thomas D. Matte, Joel Schwartz, and Richard J. Jackson, 2002, "Economic Gains Resulting from the Reduction in Children's Exposure to Lead in the United States," *Environmental Health Perspectives*, Vol. 110, pp. 563–69.

Hammitt, James K., 2010, "The Successful International Response to Stratospheric Ozone Depletion," in *Issues of the Day:100 Commentaries on Climate, Energy, the Environment, Transportation, and Public Health Policy*, edited by Ian W.H. Parry and Felicia Day (Washington: Resources for the Future).

Health Effects Institute, 2013, "Understanding the Health Effects of Ambient Ultrafine Particles," HEI Review Panel on Ultrafine Particles (Boston: Health Effects Institute).

Intergovernmental Panel on Climate Change (IPCC), 2007, *Climate Change 2007: The Physical Science Basis*, Contribution of Working Group I to the Fourth Assessment Report of the Intergovernmental Panel on Climate Change (Cambridge, U.K.: Cambridge University Press).

———, 2013, *Climate Change 2013: The Physical Science Basis*, Contribution of Working Group I to the Fifth Assessment Report of the Intergovernmental Panel on Climate Change (Cambridge, U.K.: Cambridge University Press).

International Energy Agency (IEA), 2011, *World Energy Outlook 2011* (Paris: International Energy Agency).

International Road Federation (IRF), 2009, *World Road Statistics 2009* (Geneva: International Road Federation).

———, 2012, *World Road Statistics 2012* (Geneva: International Road Federation).

Jaramillo, P., W.N. Griffin, and H.S. Matthews, 2007, "Comparative Life Cycle Air Emissions of Coal, Domestic Natural Gas, LNG, and SNG for Electricity Generation," *Environmental Science and Technology*, Vol. 41, No. 6, pp. 290–96.

Keen, Michael, Ian W.H. Parry and Jon Strand, 2013, "Ships, Planes, and Taxes," *Economic Policy*, Vol. 28, pp. 701–49.

Krewski, Daniel, Michael Jerrett, Richard T. Burnett, Renjun Ma, Edward Hughes, Yuanli Shi, Michelle C. Turner, C. Arden Pope III, George Thurston, Eugenia E. Calle, and Michael J. Thun, 2009, "Extended Follow-Up and Spatial Analysis of the American Cancer Society Study Linking Particulate Air Pollution and Mortality," Research Report 140 (Boston, Massachusetts: Health Effects Institute). http://scientificintegrityinstitute.net/Krewski052108.pdf.

Lindsey, Robin, 2006, "Do Economists Reach a Conclusion on Road Pricing? The Intellectual History of an Idea," *Econ Journal Watch*, Vol. 3, No. 2, pp. 292–379.

Mendelsohn, Robert, Roger Sedjo, and Brent Sohngen, 2010, "Forest Carbon Sequestration," in *Fiscal Policy to Mitigate Climate Change: A Guide for Policymakers*, edited by I.W.H. Parry, R. de Mooij, and M. Keen (Washington: International Monetary Fund).

Muller, Nicholas Z., and Robert Mendelsohn, 2012, *Using Marginal Damages in Environmental Policy: A Study of Air Pollution in the United States* (Washington: American Enterprise Institute).

National Research Council (NRC), 2009, *Hidden Costs of Energy: Unpriced Consequences of Energy Production and Use* (Washington: National Research Council, National Academies).

Nordhaus, William D., 2013, *The Climate Casino: Risks, Uncertainty, and Economics for a Warming World* (New Haven, Connecticut: Yale University Press).

Organization for Economic Cooperation and Development (OECD), 2013, Environmentally Related Taxes, Fees and Charges Database (Paris: Organization for Economic Cooperation and Development). http://www2.oecd.org/ecoinst/queries/index.htm.

Parry, Ian W.H., and Kenneth A. Small, 2005, "Does Britain or the United States Have the Right Gasoline Tax?" *American Economic Review*, Vol. 95, No. 4, pp. 1276–89.

———, 2009, "Should Urban Transit Subsidies Be Reduced?" *American Economic Review*, Vol. 99, pp. 700–24.

Rosen, Sherwin, 1986, "The Theory of Equalizing Differences," in *Handbook of Labor Economics* Vol. 1, edited by O. Ashenfelter and R. Layard, pp. 641–92 (New York, New York: Elsevier).

Rowlatt, Penelope, Michael Spackman, Sion Jones, Michael Jones-Lee, and Graham Loomes, 1998, *Valuation of Deaths from Air Pollution* (London: National Economic Research Associates).

Santos, Georgina, 2004, *Road Pricing: Theory and Evidence*, Research in Transportation Economics Vol. 9 (Amsterdam: Elsevier).

United States Energy Information Administration (US EIA), 2011, *International Energy Outlook 2011* (Washington: Energy Information Administration, US Department of Energy). www.eia.gov/forecasts/ieo/index.cfm.

———, 2013, International Energy Statistics (Washington: Energy Information Administration, US Department of Energy). www.eia.gov/cfapps/ipdbproject/iedindex3.cfm.

United States Environmental Protection Agency (US EPA), 2011, *The Benefits and Costs of the Clean Air Act from 1990 to 2020*, Report to Congress (Washington: US Environmental Protection Agency).

Watkiss, Paul, Steve Pye, and Mike Holland, 2005, *CAFE (Clean Air for Europe) CBA: Baseline Analysis 2000 to 2020*. Report to the European Commission (Brussels: Directorate-General for the Environment).

World Bank, 2013, World Development Indicators Database (Washington: World Bank). http://data.worldbank.org/indicator.

World Bank and State Environmental Protection Agency of China, 2007, *Cost of Pollution in China: Economic Estimates of Physical Damages* (Washington: World Bank).

World Health Organization (WHO), 2013, Global Health Observatory Data Repository (Geneva: World Health Organization). http://apps.who.int/gho/data/node.main.A997?lang=en.

Zax, Jeffrey S., and Daniel I. Rees, 2002, "IQ, Academic Performance, Environment, and Earnings," *Review of Economics and Statistics*, Vol. 84, pp. 600–16.

CHAPTER 3

Rationale for and Design of Fiscal Policy to "Get Energy Prices Right"

The first part of this chapter discusses why environmental taxes or the equivalent emissions trading systems (ETSs) should be front and center in getting energy prices right, though design details, such as targeting the right base, exploiting the fiscal dividend, and establishing stable prices aligned to environmental damage, are critical. The second part discusses a variety of further design issues, including specifics for power generation and transportation fuels, the role of other instruments, overcoming challenges to price reform, and issues for low-income countries.

Policies that emerge in practice from political processes may deviate in all sorts of ways from the economically ideal design principles outlined here. Nonetheless, having a clear sense of sound policy design helps discipline the policy debate, provides a sense of the direction in which policy should be heading, and provides a benchmark against which other, perhaps more politically palatable, policies should be evaluated to illuminate the trade-offs. And as discussed, some of the design principles carry over if regulatory approaches are chosen instead of fiscal approaches.

Other complementary policies are needed—investments in transportation and energy distribution systems, safety regulations governing the extraction and production of energy (including shale gas and nuclear) and use of roads, and so on—but are largely beyond the scope of this book, which is about pricing for residual environmental damage.[1]

POLICY INSTRUMENT CHOICE FOR ENVIRONMENTAL PROTECTION

There are three basic reasons for using fiscal instruments to address the environmental side effects of energy:

- *They are environmentally effective*—so long as they target the right base (e.g., emissions).
- *They can achieve environmental objectives at lowest economic cost*—so long as the fiscal dividend is exploited (e.g., revenues take the place of other burdensome taxes).
- *They strike the right balance between the environment and the economy*—so long as they reflect environmental damage.

---

[1]The appropriate design of these broader policies can depend on fuel pricing policies. For example, taxing coal may increase the need for power grid extensions to wind and solar generation sites.

These criteria are important, not just for their own sake, but also for credibility and sustainability. The next three subsections elaborate on these principles. Environmental taxes in a broader fiscal context, and taxes versus trading systems, are discussed in the subsequent two subsections.[2]

The Effectiveness of Alternative Environmental Policies

There are two basic points here:

First, if environmental taxes or similar pricing instruments are applied to the right base (a critical "if"), they will exploit all opportunities for reducing a particular environmental harm.

Second, in contrast, regulatory policies by themselves are typically far less effective because they are focused on a much narrower range of these opportunities—though a fairer comparison might be between pricing instruments and various combinations of regulations.

Energy-related carbon dioxide (CO_2) emissions are used to illustrate these points. The discussion goes into detail, given the importance of choosing the right instrument and that relatively ineffective instruments are often used in practice. Box 3.1 illustrates similar points about the effectiveness of well-targeted fiscal instruments in other policy contexts.

Opportunities for reducing energy-related CO_2 emissions

Opportunities for mitigating energy-related CO_2 emissions can be classified as follows:

- *Increasing the use of renewable generation fuels*—that is, shifting the power generation mix from fossil fuels to carbon-free renewables like wind, solar, and hydro.
- *Other options for reducing the emissions intensity of power generation*—including shifting from high-carbon-intensive coal to intermediate-carbon-intensive natural gas and from these fuels to carbon-free nuclear. Emissions intensity might also be reduced through adoption of carbon capture and storage technologies, provided they become viable in the future.
- *Reducing electricity demand*—through the adoption of energy-saving technologies, such as more energy-efficient lighting, air conditioners, or appliances, and reducing the use of electricity-consuming products.
- *Reducing transportation fuel demand*—by raising the average fuel efficiency (kilometers per liter) of vehicle fleets (e.g., adopting technologies to improve engine efficiency or reduce vehicle weight, shifting to smaller vehicles or various classes of electric vehicles) and curbing vehicle-kilometers driven (reducing vehicle ownership and the intensity with which vehicles are used).
- *Reducing direct fuel usage, mainly for heating, in homes and industry*—again, through adoption of energy-saving technologies such as insulation upgrades, or reduced product use such as turning down the thermostat.

---

[2]For more discussion of some of the issues covered below see, for example, Goulder and Parry (2008), Hepburn (2006), IMF (2008), Krupnick and others (2010), OECD (2010), and Prust and Simard (2004).

BOX 3.1

Environmental Effectiveness of Alternative Instruments: Further Examples

Sulfur dioxide emissions. As discussed in Chapter 4, sulfur dioxide (SO_2) emissions from coal-fired power plants are a major cause of premature death. Options for reducing these emissions include the following:

- Installing filter technologies in smokestacks to capture or "scrub" SO_2 (turning it into sludge and solid waste for impoundment in landfills or recycling)—some scrubbing technologies can capture 90 percent or more of emissions
- Shifting to coal with a lower sulfur content
- Washing coal at processing plants, which lowers sulfur content (and other impurities)
- Shifting from coal to other generation fuels (natural gas, renewables, and others) by retiring coal plants
- Reducing the demand for electricity

Charging for SO_2 emissions released from smokestacks addresses all these possibilities—the first four options lower generators' tax liabilities, while the pass-through of emissions taxes and abatement costs to consumers addresses the last response through higher electricity prices. Alternatively (and perhaps administratively easier), all these responses could be addressed by a tax levied on coal use in proportion to its emission rate, with appropriate crediting for use of emissions control technologies.

In contrast, mandating SO_2 control technologies exercises the first response, but not the others. Limiting average SO_2 emissions per unit of electricity over generators' portfolios of plants is more effective, in that it elicits the first three responses, but provides only weak incentives for the last two because generators do not pay for the full social cost of coal plants. A coal tax unrelated to pollution addresses the last two options but misses the first three.

Road traffic congestion. Various possibilities for reducing urban road congestion are available (taking transportation infrastructure as given), including encouraging people to

- Carpool
- Use alternative transport modes (bus, rail, bike, walk)
- Reduce trip frequency by telecommuting, combining several trips, or simply reducing travel
- Set off later or earlier to avoid the peak within the rush hour
- Avoid the rush hour altogether by driving off peak

Charging motorists per kilometer driven on busy roads, and varying the charge as congestion rises and falls during the course of the rush hour, gives rise to all these responses because each would reduce motorists' tax liabilities.

A simple toll per kilometer driven that does not vary with time of day is less effective because it does not encourage the last two responses. And a transit fare subsidy is far less effective still because it exploits only the second response (and even then it does not encourage shifting to other nontransit alternatives).

Effect of carbon pricing

Pricing all fossil fuel CO_2 emissions through a carbon tax (or ETS) exploits all five of the mitigation opportunities because the emissions price is reflected in higher prices for fuels and electricity.

For illustration, suppose that 25 percent of the CO_2 reduction comes from shifting to renewable generation fuels, 25 percent from other measures to reduce CO_2 per kilowatt-hour (kWh) in power generation, 20 percent from reductions in electricity demand, 15 percent from reductions in transportation fuels, and 15 percent from reductions in direct fuel consumption in homes and industry—with reductions in electricity demand and transportation fuels split equally between energy efficiency improvements and reductions in product demand.[3] These illustrative reductions are summarized in the first row of Figure 3.1, where the lengths of the green bars are scaled to the emissions reductions forthcoming from different sources.

Effectiveness of regulatory policies relative to carbon pricing

Other rows in Figure 3.1 illustrate the relative effectiveness of various alternative policies; each is scaled such that the emissions reductions for the particular source targeted by the policy are the same as those under carbon pricing.[4] Light gray bars indicate a source of emissions reduction and dark gray bars indicate policies that actually increase emissions (by lowering energy costs). Other policies typically have limited effectiveness because they fail to exercise many of the mitigation opportunities exploited under carbon pricing.

For example, a subsidy for renewable generation fuels misses 75 percent of the emissions reductions achieved by carbon pricing. And a vehicle fuel efficiency standard misses 93 percent of those reductions. In fact, fuel efficiency regulations cause a partially offsetting increase in emissions, by lowering fuel costs per kilometer and thereby increasing driving, though this "rebound effect" appears to be relatively modest.[5]

However a package of regulations can be much more effective than individual regulations. For example, a CO_2 per kWh standard will address all opportunities to lower CO_2 emissions per kWh in power generation (50 percent of the emissions

---

[3] These assumptions are based approximately on carbon price analyses for the United States in Krupnick and others (2010) and Parry, Oates, and Evans (2014). Most of the low-cost mitigation options are related to fuel switching in power generation, given the array of alternatives to high-carbon coal. In transportation, options for switching from fossil fuels to cleaner fuels remain limited, and vehicle fuel efficiency is already encouraged by high fuel prices and fuel efficiency regulations. One complication (not considered here) is the possibility that future pledges to penalize carbon emissions might, by lowering the future returns to fossil fuel production, accelerate incentives for fuel production in the nearer term, thereby undermining some of the emissions benefits. For alternative perspectives on this possibility, see Sinn (2012) and Cairns (2014).

[4] For example, an electricity tax is assumed to reduce electricity demand by the same amount as a carbon tax, and a renewables subsidy is assumed to cause the same emissions reduction (by shifting from fossil fuels for generation to renewables) as the carbon tax.

[5] The illustration assumes this effect offsets 10 percent of fuel savings from improved fuel efficiency, based on Small and Van Dender (2006).

Figure 3.1 Illustrative Sources of Fossil Fuel CO_2 Reductions under Different Policies

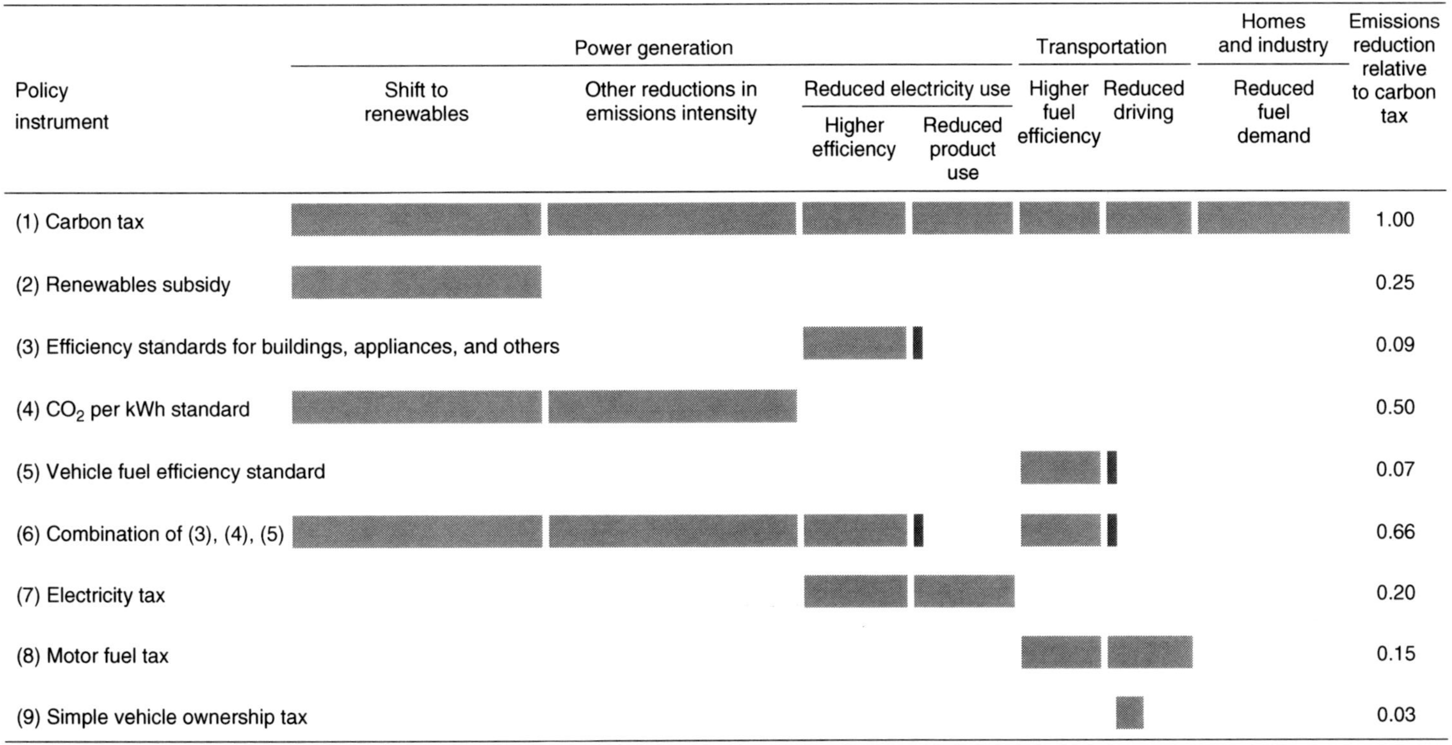

Policy instrument	Power generation		Reduced electricity use		Transportation		Homes and industry	Emissions reduction relative to carbon tax
	Shift to renewables	Other reductions in emissions intensity	Higher efficiency	Reduced product use	Higher fuel efficiency	Reduced driving	Reduced fuel demand	
(1) Carbon tax	▓	▓	▓	▓	▓	▓	▓	1.00
(2) Renewables subsidy	▓							0.25
(3) Efficiency standards for buildings, appliances, and others			▓					0.09
(4) CO_2 per kWh standard	▓	▓						0.50
(5) Vehicle fuel efficiency standard					▓	▪		0.07
(6) Combination of (3), (4), (5)	▓	▓	▓		▓	▪		0.66
(7) Electricity tax			▓	▓				0.20
(8) Motor fuel tax					▓	▓		0.15
(9) Simple vehicle ownership tax					▪			0.03

Sources: Based approximately on analyses for the United States in Krupnick and others (2010) and Parry, Evans, and Oates (2014).
Note: Light gray bars indicate emissions reductions induced by different policies; policies are scaled (when applicable) to have the same effect, for example, on fuel efficiency and electricity demand, as a carbon tax. Dark gray bars indicate sources of increased emissions because lower per unit energy costs increase demand for energy-using products (the rebound effect).

reductions under the carbon tax). If combined with comprehensive policies to improve the energy efficiency of vehicles and electricity-using equipment, such a package could exploit up to two-thirds of the emissions reductions under the carbon tax. But some mitigation opportunities are missed, such as encouraging people to drive less. Moreover, regulatory packages can be excessively costly without extensive credit trading (as discussed below), and they do not raise revenue.

Targeting the right base

Other "proxy" taxes are usually far less effective than carbon pricing (again, see Figure 3.1). For example, excise taxes on electricity consumption miss 80 percent of the mitigation opportunities achieved by carbon pricing. Also prevalent are vehicle ownership taxes (sales excises as well as registration fees and annual road charges), but these taxes, at least in their simple form unrelated to emissions, miss about 97 percent of the mitigation opportunities brought about by carbon pricing in this example.[6]

Cost Effectiveness: A First Look

Besides their effectiveness, the second main rationale for fiscal instruments is that they achieve a given environmental improvement at the lowest overall cost to the economy.

The focus of this discussion is on economic costs, as defined in Box 3.2, rather than other metrics such as GDP and employment. For now, policies are compared based on a limited definition of cost that ignores important links with the broader fiscal system (discussed in the main text below).

Environmental taxes (and ETSs) are cost-effective because they promote equalization of incremental mitigation costs across different behavioral responses. For example, with the same price on all CO_2 emissions, firms and households alike face the same incentives to alter their behavior in ways that reduce emissions up to the point at which the cost of the last ton of CO_2 reduced equals the emissions tax. ETSs are also cost-effective in this regard (so long as credit trading markets are fluid) because they also establish a uniform price on emissions across different sources.

Traditional regulatory policies, in contrast, may not perform well on cost-effectiveness grounds if they require that all firms meet the same standard and there is significant variation in pollution intensity among firms (implying it is much more costly for some to meet the standard than others). Credit trading in fluid markets is required to make these policies cost-effective, including provisions allowing the following:[7]

- Firms, such as generators with relatively emissions-intensive plants, to fall short of the standard (e.g., a limit on average emissions per kWh produced)

---

[6]This assumes (based on Fischer, Parry, and Harrington, 2007) that one-third of the reduction in driving in response to higher fuel prices comes from reduced vehicle ownership and two-thirds from reductions in kilometers driven per vehicle (a vehicle excise tax only induces the first response).

[7]Empirical studies (see, e.g., Newell and Stavins, 2003, and the references contained in their footnote 2) have documented cases of substantial cost savings from these types of flexibility provisions.

by purchasing credits from other firms that have relatively clean plants that exceed the standard
- Firms to trade credits across different regulatory programs, thereby establishing a uniform emissions price across all sectors covered by regulation.

BOX 3.2

Defining Economic Costs

The economic costs of environmental policies refer to the costs or benefits of the various ways households and firms respond to the policy, both directly (e.g., through responses to higher fuel prices) and indirectly (e.g., through the responses to broader taxes that might be reduced because of the increase in environmental tax revenues).

Staying with the carbon pricing example, the costs of the direct behavioral responses include, for example

- Higher production costs to power generators from using cleaner, but more expensive, fuels
- Costs to households from driving less than they would otherwise prefer (the value of trips forgone, less savings in time and fuel costs)
- Costs to firms and households from switching to more energy-efficient vehicles, appliances, machinery, and so on—that is, the upfront purchase costs less the life cycle savings in fuel costs.

More generally, higher energy and transportation costs tend to slightly contract the overall level of economic activity, which, in turn, may slightly reduce economy-wide employment and investment. As discussed in the main text below, employment and investment levels are already distorted by taxes on work effort and capital accumulation, and environmental policies cause economic costs if they worsen these distortions. However, environmental tax revenues provide an offsetting benefit if they are used to lower the taxes on work effort and capital accumulation.

Economic costs do not include pure dollar transfers between the private sector and the government—whatever dollar amount the private sector pays in tax liabilities is offset by a revenue benefit to the government. Economic costs are also quite different from job losses in industries burdened by environmental policies (at least some of which are made up by other sectors after a, perhaps lengthy, adjustment period)—employment effects do matter for costs, as just noted, but the issue is entwined with the employment effects of the broader tax system. Economic costs need not be closely related to changes in GDP either (Krupnick and others, 2010, p. 23).

The concept of economic costs has been endorsed by governments around the world for purposes of evaluating government spending, tax, and regulatory policies. In the United States, for example, a series of executive orders since the 1970s has required government agencies to conduct hundreds of cost-benefit assessments a year, based on the notion of economic costs, to determine whether major policy initiatives are warranted from society's perspective.

> The broader macroeconomic impacts of environmental taxes are likely to be modest, at least if revenues are used to lower burdensome taxes on work effort and capital accumulation. The short-term macroeconomic impacts could be larger if revenues are used to reduce budget deficits, though that would be true of any fiscal consolidation measure. The fiscal crisis late in the first decade of the 2000s should not detract from efforts to price environmental damage, not least because of their contribution to badly needed revenues (Jones and Keen, 2011).

Balancing Benefits and Costs

The third attraction of tax and pricing policies—that they can strike the right balance between environmental benefits and costs to the economy—requires that prices be set equal to their "corrective" levels, that is, equal to incremental environmental damage. If prices are less than environmental damage, some socially desirable environmental improvements will be forgone; if prices exceed environmental damage, some environmental improvements will be made that are not justified by their cost.[8]

The corrective tax calculations in Chapter 6 assume that the environmental damage per unit is constant (e.g., damage per ton of emissions, or congestion costs per vehicle-kilometer traveled). This seems reasonable, for example, for CO_2 damage, because one country's emissions in one year add very little to the atmospheric accumulation of greenhouse gases. The assumption might appear more questionable for local air pollution, implying, as it does, that the corrective tax should be independent of local ambient air pollution, though some justification is provided in Box 3.3.

Constant environmental damage per unit also favors the use of instruments that price emissions (but that allow the quantity to vary with changes in energy demand, fuel prices, and so on) over instruments that fix the quantity of emissions but allow prices to vary. The latter are more appropriate when the thresholds beyond which environmental damage rises sharply are known (Weitzman, 1974), though these cases do not seem especially relevant for this volume.

Damage per unit also varies across regions within a country, for example, the human health effects of air pollution depend on local population exposure. In principle, charges could be differentiated according to the location of the emissions source, though this complicates administration—mostly because pollution from tall smokestacks can be transported great distances (Chapter 4). Moreover, studies (e.g., Muller and Mendelsohn, 2009) suggest that imposing an appropriately scaled, uniform emissions charge produces the biggest net benefit—differentiating the charge according to region produces additional, but smaller, benefits.

---

[8] Corrective taxes are sometimes called Pigouvian taxes after the economist Arthur Pigou, who first recommended them.

BOX 3.3

Shape of the Air Pollution Damage Function

As discussed in Chapter 4, some evidence suggests that the relationship between environmental damage and ambient air pollution concentrations begins to flatten out at higher levels of pollution concentration (because people's ability to take in more pollution becomes saturated). Therefore, the slope of the environmental damage function is flatter at the high pollution concentration C_1 in Figure 3.3.1, compared with the slope at the lower concentration level C_0. Thus, additional pollution emissions do less harm at concentration C_1 than at C_0.

Figure 3.3.1 Shape of the Air Pollution Damage Function

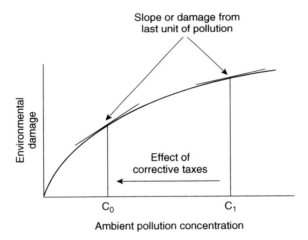

This function might suggest that, given other factors, the corrective tax on fuel or emissions should be lower, paradoxically, in high-pollution countries. This possibility is ignored in this study, however. If corrective taxes of the scale typically estimated in this volume were to be introduced, emissions would fall dramatically, most likely lowering pollution concentrations below levels at which the damage curve might flatten out.

Environmental Taxes in a Broader Fiscal Context

This subsection discusses the distortions to economic activity created by the broader fiscal system and how environmental taxes impact these distortions. The appropriate treatment of energy products under value-added taxes (VAT) or similar sales tax systems is discussed in Box 3.4, but, assuming normal procedures are applied, is not relevant for corrective tax design.

BOX 3.4

Coverage of Energy Products under the Value-Added Tax (VAT)

Leaving aside environmental considerations, basic tax principles suggest that all consumption goods should be included under any VAT (or other general sales tax) system, to raise revenues without distorting choices among consumer goods. Intermediate inputs should be excluded from VAT to avoid distorting firms' choices about the mix of labor, capital, energy, and material inputs. These principles apply automatically under normal VAT systems, so long as intermediate goods are sold to entities paying VAT—any taxes paid under the VAT for power generation fuels and electricity used by industry are, appropriately, reimbursed.

In contrast, any tax rate applying to consumer goods in general should also be applied to household purchases of electricity, vehicles, and fuels. Ideally, any VAT should apply to fuel prices including corrective taxes, to avoid distorting choices among consumption goods, taking account of their full social costs of production. Corrective fuel tax estimates reported in Chapter 6 are before any application of VAT.

Taxes on labor income (e.g., personal income and payroll taxes) can cause large differences between wages that workers take home and compensation paid by firms. As a result, these taxes tend to depress work effort by discouraging labor force participation, overtime, effort on the job, investments in human capital, and so on. The VAT (and other consumption taxes) has similar effects because it reduces the amount of goods that can be purchased from a given amount of work effort.

Similarly, taxes on corporate income and individual income from savings tend to burden the economy by reducing capital accumulation below the level that would otherwise occur.

Environmental taxes (or similar instruments) interact with these sorts of distortions in two opposing ways (Goulder, 2002; Parry and Oates, 2000).[9]

First, because environmental taxes are passed forward into the prices for fuels, electricity, transportation, and so on, they tend to contract, albeit very slightly, the overall level of economic activity, which, in turn, reduces employment and investment. In effect, energy taxes act like implicit taxes on labor and capital, and they compound the adverse effects of those taxes. Studies find that the costs of environmental taxes (or ETSs) are considerably higher—possibly several times higher (Goulder, 2002; Parry and Oates, 1999)—when their adverse effects in tax-distorted factor markets are taken into account.

---

[9] This discussion abstracts from the possibility of other, nontax, distortions in labor markets, such as wage rigidity (e.g., from union power) and chronically deficient demand.

Second, however, if environmental tax revenues fund broader tax reductions, relatively large economic benefits are produced by increasing incentives for work effort and capital accumulation, reducing incentives for informal activity, and reducing distortions from tax preferences (e.g., for housing). This offsets most, or more than offsets the adverse effects on factor markets from higher energy prices, implying moderate costs overall for the economy and perhaps negative costs if revenues are used to reduce a particularly distortive tax. Although environmental tax revenues are earmarked more often than other taxes, there are notable examples of environmental tax revenues substituting for other taxes (Box 3.5). Of course, there are alternative revenue uses that might yield comparable economic benefits, including reducing budget deficits, which, in turn, lowers future tax burdens, and funding socially desirable spending.

Despite the fiscal dividend, this is not necessarily a reason to set higher tax levels—roughly speaking, environmental taxes should be set on environmental grounds, with further revenue requirements met through broader fiscal instruments (personal income taxes, VAT, and so forth).

BOX 3.5

Environmental Tax Shifting in Practice

Several countries have introduced, or increased, environment-related taxes while simultaneously cutting other taxes:

- In the early 1990s, Sweden introduced taxes on oil and natural gas to charge for carbon and (for oil) sulfur dioxide and on coal-related sulfur dioxide and industrial nitrogen oxide emissions. These reforms were part of a broader tax-shifting operation that also strengthened the value-added taxes while reducing taxes on labor and traditional energy taxes (on motor fuels and other oil products).
- Between 1999 and 2003, Germany increased taxes on transportation fuels and introduced new taxes on natural gas, heating fuels, heavy fuel oil, and primarily residential electricity consumption. About 85 percent of the revenue was used to fund reductions in employer and employee payroll taxes, about 14 percent was used for budget consolidation, and 1 percent for renewable energy programs.
- In Australia's carbon pricing scheme, which covers about 60 percent of carbon dioxide emissions, about three-quarters of the allowances are auctioned, with half of this revenue used to fund a tripling of personal income tax thresholds. This scheme was implemented in 2012, but now looks likely to be scrapped.
- From 2008 to 2012, British Columbia progressively introduced a carbon tax covering 70 percent of fossil fuel emissions, with more than 90 percent of the revenues used to fund reductions in personal and corporate taxes.

Sources: Department of Climate Change and Energy Efficiency (2011); Government of British Columbia (2012, pp. 66).

However, insofar as possible, exploiting the fiscal dividend is critical, given the large amount of revenue at stake (Chapter 6). If revenues from environmental taxes are not used productively, the overall costs of environmental taxes can be substantially higher.[10]

This last point has some notable policy implications:

- *Be wary of earmarking* environmental tax revenues, such as for clean energy programs or climate adaptation. Ideally, any earmarked spending should generate comparable economic benefits from alternative revenue uses (e.g., lowering other tax burdens).[11]
- *Compensate appropriately*: Compensation payments (e.g., for groups especially vulnerable to higher energy prices) may have high equity and political value, but can significantly reduce the overall cost-effectiveness of environmental taxes by reducing revenues for other, perhaps more economically efficient, purposes such as cutting other burdensome taxes. Policymakers need to evaluate the trade-offs carefully.
- *If possible, use compensation schemes that improve economic efficiency:* Tensions between compensation and cost-effectiveness might be ameliorated if the compensation scheme produces economic benefits. For example, providing relief to low-income households through tax reductions (e.g., lowering the basic rate of tax, rebating payroll taxes, providing earned income tax credits) improves incentives for work effort, whereas transfer payments made regardless of work effort do not.

Taxes versus ETS: A Quick Look

In principle, the choice between emissions taxes and ETS is less important than implementing one of them and getting the design details right. The most important of these details are the following:

- Comprehensively covering the sources of environmental harm
- Prudently using the fiscal dividend
- Scaling program stringency to environmental damage
- Establishing stable and predictable prices.

---

[10]In fact, environmental taxes can lose their cost-effectiveness advantage over regulatory alternatives (Goulder and others, 1999; Parry and Williams, 2012). Environmental taxes can cause larger increases in energy prices (because they involve the pass-through of tax revenues in higher prices) than regulatory policies (which do not raise revenues): the larger the increase in energy prices, the larger the adverse effects on tax-distorted factor markets—and hence the need to offset these higher costs with productive use of revenues.

[11]Another problem with earmarking is that there is no relationship between the economically appropriate amount of spending on clean energy or related programs and the amount of revenues raised by a corrective environmental tax. The tendency with earmarked revenues has been to set tax levels to meet spending requirements, resulting in tax rates frequently well below levels needed to correct for environmental damage (Opschoor and Vos, 1989).

Figure 3.2 Price Experience in the European Union Emissions Trading System

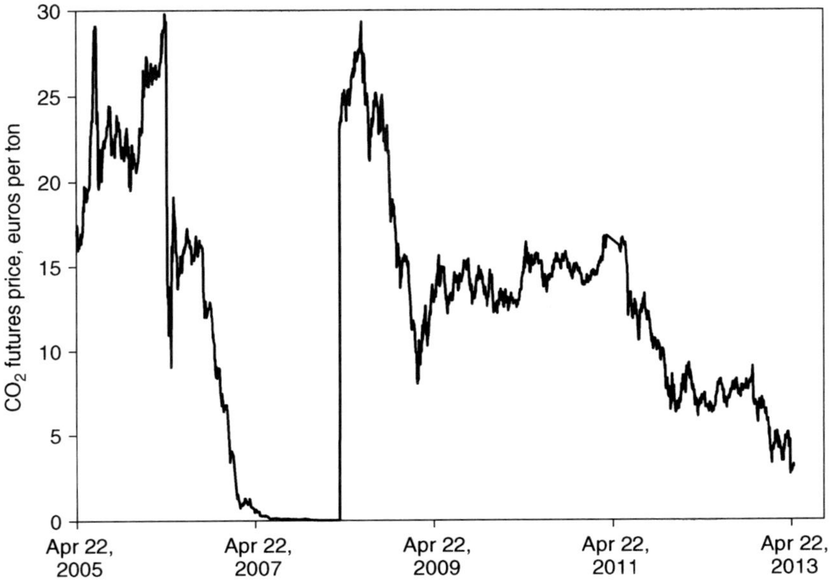

Source: Bloomberg (2013) EU-ETS futures series "MO1 Comdty."
Note: CO_2 = carbon dioxide.

The importance of the first three design features was discussed above.

The fourth helps to contain program costs. Pure ETS systems carry the danger that either the cap will be too low at a given time (e.g., because energy demand turns out to be lower than expected), in which case emissions prices are depressed and low-cost mitigation opportunities forgone, or the cap will be too tight, such as in periods of high energy demand, in which case emissions prices and mitigation costs are excessive. A predictable and rising emissions price also fosters expectations that policy will be sustained, which is important for clean technology development and deployment (especially technologies with high upfront costs and long-range returns). Predictable prices also reduce uncertainty about revenues. Price volatility has been a problem in some trading systems, for example, in the European Union where futures prices reached €30/ton of CO_2 in 2006 and 2008, but fell to less than €5/ton in 2007 and again in 2013 (Figure 3.2).

Striving for price stability implies a political trade-off, however, given that emissions vary from year to year—to date, climate policy goals have usually been expressed as annual emissions levels rather than price targets.[12]

---

[12]"Carbon budgets" are a possible compromise between emissions targets and price targets. Carbon budgets would specify allowable emissions cumulated over a long period (say 10 years), but allow year-to-year variability in emissions. They could be implemented through a carbon tax whose rate might be adjusted later in the budget period if cumulative emissions would not otherwise be on track to stay within budget.

Appropriately designed environmental taxes naturally meet the four design features. So do ETSs if allowances are auctioned and price stability provisions such as price floors and ceilings are included.[13]

In practice, a carbon tax seems to be the instrument more likely to fully exploit the fiscal dividend because it would presumably be administered by a finance ministry, for example, as an extension of existing motor fuel excises to other petroleum products, coal, and natural gas. An environmental agency administering an ETS might be reluctant or legally unable to remit all revenues from allowance auctions to the finance ministry.

Tax and pure ETS systems can also interact differently with other climate-related policies. For example, policies to promote renewables and energy efficiency will not affect emissions if emissions are held fixed by a cap—instead, they lower emissions prices (if there is no price floor) and the potential fiscal dividend from an ETS.[14] In the presence of carbon taxes, however, other policies reduce emissions (rather than the emissions price, which is fixed by the tax) and therefore tend to have a much weaker impact on reducing revenues.

It has been suggested that an ETS is the more natural instrument for channeling climate finance to developing countries because it can be combined with offset programs. Offsets could be included under a tax, however—tax credits could be awarded to domestic fuel suppliers for funding climate mitigation projects in developing countries. Again, offset provisions, assuming they promote genuine emissions reductions as opposed to reductions that would have occurred anyway, improve environmental effectiveness under a tax without substantially reducing revenues. But under an ETS with no price floor, offset provisions lower the emissions price and carbon pricing revenues with no effect on total emissions.

FURTHER DESIGN ISSUES

Several complicating factors are now considered, relating to overlapping environmental problems, the interactions with and roles of other policies, measures to improve the acceptability of policy reform, and relevance for low-income countries. Further issues are touched on in Box 3.6.

Multiple Environmental Problems

As discussed in Chapter 2, energy and transportation systems cause more than one source of environmental harm—fuel combustion leads to both carbon

---

[13]Studies for the United States (Fell, MacKenzie, and Pizer, 2008) suggest that an ETS scaled to achieve the same cumulative emissions reduction as a carbon tax would be moderately (about 15 percent) more costly over time if the ETS lacked price stability provisions.

[14]In practice, however, policymakers may be more willing to tighten an emissions cap if emissions prices are reduced through other policies.

BOX 3.6

Unintended Consequences and Market Price Distortions

This box discusses examples of unintended consequences from tax reform and the implications of market price distortions. An illustration of unintended consequences is when environmental taxes cause increased use of substitute fuels with environmentally harmful effects that cannot themselves be priced.

For example, fossil fuel taxes can lead to excessive reliance on nuclear power, with its consequent risks of meltdowns and unsafe waste disposal. In the absence of effective regulatory and liability frameworks to adequately minimize these risks (they are difficult to address through taxes on nuclear power, primarily because of uncertainty about the probability and scale of disasters), taxes on fossil fuels for generation should be phased in more cautiously. Similar issues arise if heavy taxation of coal use leads to extraction of domestic shale gas reserves, which, without safeguards, could pose significant environmental risks (e.g., from groundwater pollution or escape of methane—a potent greenhouse gas).

To take another very specific example, fine particulate concentrations are especially high in Ulaanbaatar, Mongolia as the result of coal use for winter home heating. However, if residential coal prices were significantly increased through taxation, households may be induced to instead burn highly toxic plastics, rubber, or other difficult-to-price garbage, with possibly even worse consequences for human health (World Bank, 2011). The problem is the lack of viable cleaner alternative fuels—until such alternatives, perhaps imported oil or liquefied natural gas, are provided, high residential coal taxes may not make sense (although taxation of coal for power generation should not be precluded). In principle, the corrective fuel tax in this instance would be adjusted downward by the increase in the use of other fuels per unit decrease in the taxed fuel, times the per unit environmental damage from the other fuel.

Another example is related to tax competition. Higher diesel fuel taxes in one European country could induce truckers to refuel in neighboring countries with lower tax rates, offsetting some of the domestic pollution and congestion benefits. One response is for countries to coordinate their minimum fuel tax rates (T&E, 2011). Another is to partially transition to kilometer-based tolls that truckers must pay regardless of where they purchase fuel.

Environmental taxes can also interact with market price distortions. If, because of limited competition, product prices exceed per unit production costs, consumption of the product will be too low from an economic perspective. In principle, a lower environmental tax affecting such an industry might be called for, though in practice any adjustments may be very modest (Oates and Strassmann, 1984). In other cases, such as at state-owned enterprises, product prices may fall short of per unit production costs, though whether this justifies tax levels significantly higher than environmental damage is an open question (especially because of lack of transparency about the extent of price distortions). In either of these cases, however, the ideal policy would be to remove the market price distortion and set the environmental tax on environmental grounds.

emissions and local air pollutants, while broader adverse side effects of vehicle use include, for example, road congestion. The implications for fiscal policy design are discussed below, taking in turn power generation, heating, and transportation fuels.

Power generation fuels

Fossil fuel combustion at power plants causes carbon emissions and local air pollutants, the most important of which is fine particulates (particulate matter with diameter of up to 2.5 micrometers, or $PM_{2.5}$). These particulates are emitted directly (e.g., during coal combustion) and are formed indirectly from chemical transformations of sulfur dioxide (SO_2) in the atmosphere, and to a lesser extent (because it is less reactive), nitrogen oxides (NO_x). SO_2 is primarily caused by coal combustion, and NO_x is produced from all fossil fuels.

Damage from these emissions is generally additive (with some caveats—see Chapter 4) and can be addressed either through taxes imposed on fuel supply in proportion to emissions factors (with appropriate crediting for any emissions capture at the point of fuel combustion) or taxes imposed directly on emissions.

For coal, the appropriate set of charges on fuel input at power plants includes the following four charges:

- Tons of CO_2 per unit of energy times damage per ton of CO_2
- Tons of $PM_{2.5}$ per unit of energy times damage per ton of $PM_{2.5}$
- Tons of SO_2 per unit of energy times damage per ton of SO_2
- Tons of NO_x per unit of energy times damage per ton of NO_x

Emission rates are defined with respect to energy rather than tons of coal given the variation in energy content across different coal types (discussed in Chapter 4). Default charges could reflect emission rates before any application of emissions control technologies—for which reasonable data are generally available by country from independent sources (Chapter 4)—with the onus on the plant operator to demonstrate, through use of continuous emissions monitoring technologies, any reduction in emission rates from application of control technologies to receive an appropriate tax credit.

Alternatively, if charges are levied on emissions rather than fuel input, the appropriate rate per ton of emissions would simply reflect environmental damage per ton, with separate charges applicable to each of the four pollutants. This approach might be more complex to monitor and enforce because governments need to assemble local air emissions data themselves by ensuring that all plants install, and correctly operate, continuous emissions monitoring technologies.

The same design principles apply to natural gas plants, though emissions damage is much lower, and charges for primary particulates and SO_2 may not be needed because the emission rates for these pollutants are, at most, very small.

Heating fuels

A fuel charge is likely the preferred regime for heating fuels such as natural gas, given the large number of small-scale emissions sources in the household sector. Again, the same principles for aggregating over emissions sources (though perhaps for just CO_2 and NO_x) as just discussed would apply.

Transportation

Fuel taxes are discussed first (the focus of this volume) followed by kilometer-based taxes, which are a possibility for the longer term.

Motor fuel taxes: Consider first gasoline taxes for passenger vehicles, for which there are four main environmental side effects: CO_2, local air pollution, traffic congestion, and traffic accidents.[15]

If fuel taxes are the only fiscal instrument available, the corrective tax, expressed in fuel units following normal practice, is given by Parry and Small (2005):

$$[CO_2 \text{ damage per liter}] \quad (3.1)$$
$$+$$
[(congestion, accident, and local pollution costs imposed on others per extra kilometer of driving)
$$\times$$
(kilometers per liter)
$$\times$$
(fraction of the fuel reduction resulting from reduced driving rather than from higher fuel efficiency)]

The first component in this formula charges for CO_2 emissions and is calculated by CO_2 emissions per liter—essentially the same for gasoline across all countries[16]—times CO_2 damage per ton.

The second component reflects effects varying with vehicle-kilometers—congestion, accidents, and local pollution (see below)—rather than fuel use. Multiplying their combined, economy-wide average costs per kilometer by fuel efficiency (kilometer per liter) expresses the effects in costs per liter.[17]

Distance-related costs are also multiplied by the fraction of the tax-induced reduction in gasoline use resulting from reduced driving, as opposed to the

---

[15] For further discussion of these side effects, see de Borger and Proost (2001); CE Delft, Infras, and Faaunhofer ISI (2011); Delucchi (2000); Maibach and others (2008); Quinet (2004); Santos and others (2010); US FHWA (1997).

[16] This discussion focuses on the taxation of pure gasoline and leaves aside the taxation of ethanol (which is sometimes blended with gasoline) and compressed natural gas, both of which are relevant for specific countries.

[17] Ideally, account should be taken of how fuel efficiency responds to changes in fuel taxes, though for simplicity the calculations below omit this complication.

fraction resulting from fuel efficiency improvements. (As a rough rule of thumb this fraction is 0.5.) The smaller this first fraction, the smaller the benefits per liter of fuel for reduced congestion, accident, and local pollution, implying a smaller corrective fuel tax would be appropriate.

Note that in the second component in equation (3.1) costs could alternatively be expressed per unit of fuel use, rather than per kilometer, skipping the need to measure fuel efficiency (for which accurate data are often lacking)—so long as these costs are scaled by the driving fraction of the fuel response.

For gasoline (as for natural gas) the major local air pollutant is NO_x. Equation (3.1) assumes that reductions in kilometers driven reduce these emissions, while fuel efficiency improvements do not. This assumption is reasonable in countries such as the United States, where the same emissions per kilometer (or mile) standards are imposed on all vehicles, irrespective of their fuel efficiency, and emission rates are maintained, at least to some extent, throughout the vehicle life by inspections programs (Fischer, Parry, and Harrington, 2007). In countries lacking these regulations, local emissions would instead be proportional to fuel use, and would therefore enter the first component in equation (3.1).

Equation (3.1) applies to motor diesel used by commercial trucks, but with some modifications (aside from different input values). Local air emissions from trucks have not traditionally been regulated on an emissions per kilometer basis and are therefore proportional to fuel use (i.e., they should not be multiplied by the driving portion of the fuel demand response). In addition, trucks are almost entirely responsible for vehicle-induced wear and tear on roads given that road damage is a rapidly escalating function of a vehicle's axle weight (Evans, Winston, and Small, 1989), and this damage should enter the second component of the corrective fuel tax.[18]

In practice motor diesel is used by both cars and trucks. Given administrative complications in differentiating the price at the pump paid by these different vehicles, the main corrective tax estimates presented in this analysis average car and truck fuel use, though it turns out not much differentiation would be warranted anyway. However, a much lower tax rate would be appropriate for off-road uses of diesel, such as in farm and construction vehicles, though differentiation in this case is more feasible (e.g., by putting dyes into fuels).

Distance-based taxes: In the longer term, countries ideally should be partially shifting away from fuel taxes toward kilometer-based taxes to more effectively address side effects that vary directly with distance traveled (Johnson, Leicester, and Stoye, 2012).

For congestion, per kilometer tolls on busy roads that progressively rise and fall during the course of the rush hour exploit all possibilities, given existing transportation infrastructure, for inducing behavioral changes to alleviate congestion (see Box 3.1).

For accidents, per kilometer charges might be scaled according to driver risks (e.g., higher for those with higher insurance company rating factors based on age,

---

[18] Noise is sometimes discussed as another adverse side effect from trucks, though is ignored here because the damage costs appear modest relative to other effects (US FHWA, 2000).

prior crash records, and the like) and perhaps would be higher for larger vehicles posing greater crash risks to others.[19] Even for local air emissions, a better corrective tax might be a per kilometer toll, the rate for which depends on the emissions characteristics of the vehicle and on local population exposure to those emissions.

Finally, road damage is most efficiently addressed through per kilometer tolls on heavy trucks, scaled by their axle weight, which would encourage truckers to seek vehicle fleets that carry goods efficiently over more axles.

Thus, in principle, the ideal fiscal system for motor vehicle transport would involve charging motorists for each kilometer driven, and the charge would be scaled according to factors affecting the congestion, accident, local pollution, and possible road damage costs imposed on others by that kilometer driven, and a fuel tax component would be retained to address carbon emissions. Developments in metering technologies such as global positioning systems (GPS) suggest that kilometer-based tax systems are now feasible (see Box 3.7). The trade-offs between policy effectiveness and administration costs need careful study, however—for example, extra administration required to fine-tune local kilometer charges to emission rates and population exposure may not be worthwhile if emissions are being controlled effectively by direct regulations.

BOX 3.7

Examples of Distance-Based Charging for Vehicles

This box describes some examples of per kilometer charging systems; only the last is a nationwide scheme.

Singapore introduced an area license (or day-pass) scheme in 1975 that dramatically raised travel speeds within the restricted zone, though also initially increased congestion outside of the zone (Santos, 2005). In 1998 area licensing was replaced with an electronically debited toll on certain links, with the objective of maintaining average speeds of 30–40 miles per hour on expressways and 12–18 miles per hour on major roads. Charges rise and fall in 30-minute steps during peak periods, based on congestion levels observed in the previous quarter.

Norway experimented with cordon tolling, though with little effect on congestion given that the objective was to raise a modest amount of transportation revenue rather than deter congestion.

London introduced an area licensing scheme in 2003 with a daily congestion charge on weekdays of £8. Collection is enabled by video cameras at checkpoints into and within the priced area that record each vehicle's license plate. Penalties are mailed to drivers who have not prepaid. In the first two years, congestion fell 30 percent within

---

[19]Distance-based charges for congestion and accidents should apply equally to similar-size vehicles, regardless of their fuel type or fuel efficiency (i.e., besides traditional fuel vehicles they should also cover hybrid, electric, and natural gas vehicles).

the priced zone, though by 2008 average speeds had fallen back to previous levels because of more traffic from exempt vehicles and because some roads were reserved for the exclusive use of buses, pedestrians, and cyclists (Transport for London, 2008). A similar scheme now operates in Milan, with a fixed daily charge of €5.

Stockholm implemented a cordon toll in 2007 covering an area of about 36 square kilometers (again, enforcement is based on license plate recognition). Fees for passing the cordon vary between 10 and 20 Swedish krona with time of day, though some vehicles are exempted (emergency vehicles, buses, motorcycles, alternative fuel vehicles). Congestion initially dropped by 50 percent on the main routes approaching the city center, and 20 percent within the city center, though it subsequently deteriorated (Eliasson, 2009).

Congestion pricing is gaining limited momentum in the United States, with federal funding for pilot schemes under the Value Pricing Program and the reduction of regulatory obstacles to freeway pricing. Some schemes open up links previously reserved for high-occupancy vehicles to single-occupant vehicles in exchange for a fee (e.g., Interstate 15 in San Diego, California) while others use tolls to fund new infrastructure (e.g., lanes added to State Route 91 in Orange County, California, in 1995).

Germany introduced a nationwide tolling system (metered by GPS) for highway use by trucks weighing 12 tons or more in 2005. Charges vary between €0.14 and €0.20 per kilometer according to vehicle type, number of axles, and emission rates—eventually, the tolls will also vary with time of day and region.

Given that widespread applications of kilometer-based taxes appear to be a long way off,[20] in the interim it is still appropriate (it produces significant net economic benefits) to reflect all adverse side effects of vehicle use in fuel taxes.

Implications of Other Policies for Environmental Tax Design

The discussion now turns to a variety of other commonly implemented, mostly regulatory, policies aimed at reducing fossil fuel emissions and that may have implications for the impact or design of environmental taxes.[21]

Renewables and energy efficiency regulations: Renewables and energy efficiency regulations (for the power sector) do not affect environmental damage per unit of coal or natural gas and therefore do not change corrective taxes for these fuels. Their effect is to reduce environmental benefits from tax reform because some of the potential behavioral responses (shifting from fossil fuels to renewables, adoption of electricity-saving technologies) will already have been induced by the regulation.

Emissions trading systems (ETSs): Taking the example of a carbon tax, if the tax is targeted to the same base to which an ETS (without price floors) already

---

[20]For example, proposals for nationwide distance-based taxes in the Netherlands and the United Kingdom have been put on hold.
[21]Subsidies for fossil fuel consumption are not considered here because they act like negative fuel taxes (though existing subsidies are factored into estimates of the impacts of policy reform in Chapter 6).

applies, the tax will reduce allowance prices rather than affect emissions (which are fixed by the cap). On net, government revenues from the tax and the ETS should be unaffected if allowances are auctioned, or increased if allowances are freely allocated. In the latter case, the tax appropriates rents that would otherwise have accrued to allowance holders.[22]

Air emissions regulations: Regulations may include requirements that new plants incorporate control technologies, such as flue gas desulfurization technologies, or maximum allowable rates for emissions per kWh averaged across generators' plants (see Box 3.1). Again, these policies do not affect the appropriate charge per ton for the remaining emissions, though they do affect the appropriate tax credit needed to reflect differences between uncontrolled and controlled emission rates in a fuel charge system. They also shave off some of the effectiveness of emissions charges.

Vehicle fuel efficiency and emissions standards: Fuel efficiency standards also dampen some of the environmental effectiveness of fuel taxes. In addition, they can raise corrective motor fuel taxes by increasing the fraction of a given tax-induced reduction in fuel use that comes from reduced driving, thereby multiplying the contribution of kilometer-related costs to the corrective fuel tax (based on equation (3.1)). Emissions per kilometer standards (applied to new vehicles) have the opposite effect: by reducing average emission rates they lower the local pollution component of the corrective fuel tax.

Other Policies to Address the Limits of Environmental Taxes

Regulatory policies could potentially complement environmental taxes, for example, if practical constraints reduce the ability to implement pricing reforms, or if the policies promote responses beyond the reach of fuel taxes (e.g., emissions regulations encourage vehicle manufacturers to reduce emissions per liter, whereas fuel taxes do not). This subsection briefly discusses the design of these policies, focusing first on traditional regulations and then on more novel policies.

Traditional regulatory approaches

Desirable features of regulatory policies include the following:
- Broad coverage to promote the widest possible range of opportunities for reducing environmental harm. As mentioned earlier, for example, a CO_2 per kWh standard for power generation is much more effective than a renewables policy, because the emissions standard addresses all fuel-switching possibilities to reduce emissions rather than just a shift to renewables.
- Credit trading provisions to allow some firms to fall short of the standard, especially if meeting the standard is prohibitively costly, by purchasing credits from other firms that exceed the standard.

---

[22]This occurs in the European Union ETS because the U.K. government collects revenue from its carbon tax floor partly at the expense of allowance holders elsewhere in the European Union.

- Price ceilings and floors (despite their tendency to reduce the urgency of credit trading provisions). The ceiling allows firms to pay fees instead of fully meeting the standard, which they might do in periods when compliance costs are relatively high, while the floor allows firms to receive subsidies if they exceed the standard in periods when compliance costs are relatively low. Ideally, these price ceilings and floors would be harmonized across different regulations and they would be set so implicit prices on emissions are in line with estimated environmental damage.

These price-stability features would make regulations look more like corrective environmental taxes. However, regulations would still differ from taxes in that they do not exploit all emissions mitigation opportunities, and they do not, on average, increase revenue. They might be more politically acceptable, however, because they have smaller impacts on energy prices given that they do not involve the pass-through of taxes in higher prices.

Novel alternatives to environmental taxes

More novel options for mimicking the effects of environmental taxes, again without a large, politically difficult increase in energy or product prices, are considered below.

"Feebates" are a combination of fees and rebates. Such policies have mainly been discussed as an alternative to CO_2 emissions per kilometer (or equivalently, fuel efficiency) standards on new vehicles (Small, 2010). Feebates would involve a fee on new vehicles with above-average CO_2 per kilometer and a rebate to vehicles with below-average CO_2 per kilometer, and fees and rebates are levied in proportion to the difference between the vehicle's CO_2 per kilometer and some "pivot point." If the pivot point is the average CO_2 per kilometer of, say, last year's new vehicle fleet, the policy would be approximately revenue neutral. If revenue from vehicles is a priority, perhaps because of constraints on broader fiscal instruments, feebates can be combined with vehicle excise taxes (see Box 3.8).

BOX 3.8

Reconciling Fiscal and Environmental Objectives in Vehicle Taxation

Vehicle excise taxes are often related to CO_2 per kilometer, with vehicles classified into different brackets and more favorable taxes applied to the lower emission rate brackets. These excise taxes are an improvement over tax systems related to engine capacity because they address some emissions-saving opportunities (e.g., reducing vehicle weight or improving rolling resistance) that the latter do not.

But one problem with these schemes is that they set up a tension between revenue and environmental objectives—the more successful the policy in shifting people to lower emissions vehicles, the lower the tax receipts. Moreover, tax brackets do not provide ongoing incentives for manufacturers to reduce the emission rate of the vehicle once it has fallen into the next lower tax bracket.

Both problems are avoided by combining an ad valorem tax on vehicle sales with a feebate. The tax provides a stable source of revenue that does not decline as emission rates fall, and it does not distort the choice among vehicles because all vehicle prices rise in the same proportion. In addition, the feebate provides ongoing rewards for all opportunities to reduce emission rates for all vehicles.

Feebates might also be used to reduce the emissions intensity of power generation. Generators with high emissions intensity would pay fees in proportion to the difference between their emissions per kWh (averaged over their portfolio of plants) and a pivot point emissions per kWh, while generators with low emissions intensity would receive corresponding rebates.

Feebates have several attractive features, although regulations can have similar merits if accompanied by design features such as price ceilings and floors.

First, feebates are cost-effective because all firms face the same rewards for reducing emissions. Second, feebates automatically provide ongoing incentives to continually reduce emissions, whereas traditional regulations do not, because once firms have met the standard they have no incentives to exceed it. Third, fees and rebates can be set such that the implicit reward for reducing emissions approximately reflects environmental damage. Fourth, they create some winners (those receiving subsidies) in the affected industry, which could help with acceptability.

Another novel policy—one that encourages people to drive less (a response that is difficult to regulate) and without a politically difficult increase in tax burden for the average motorist—is to change automobile insurance from lump-sum payments into payments proportional to kilometers driven. This possibility is discussed in Box 3.9.

BOX 3.9

Pay-as-You-Drive Auto Insurance

One promising way to reduce vehicle-kilometers traveled in countries with well-established automobile insurance systems—but in which premiums take the form of lump-sum annual payments—is to transition to pay-as-you-drive (PAYD) insurance, under which premiums vary in proportion to the policyholder's annual kilometers.[1] Existing rating factors, as determined by insurance companies, would be used to set per kilometer charges for different drivers: inexperienced drivers, or those with prior crash records, for example, would pay higher per kilometer charges. This approach would maximize the road safety benefits because those with the greatest crash risks would have the greatest incentives to drive less.

The transition to PAYD could occur on a voluntary basis, with the government kick-starting the process using tax incentives.[2] Drivers with below-average annual kilometers would have the strongest incentives to take up PAYD (under the current system, low-kilometer drivers subsidize high-kilometer drivers) and as they switched, premiums would rise (to maintain insurance company profits) for the remaining pool of drivers with lump-sum insurance, encouraging further shifting to PAYD. Global positioning systems and nearly tamperproof odometers (with appropriate safeguards) now provide a potentially reliable and accurate way to collect information on kilometers driven.

---

[1] Existing systems often provide a modest discount for drivers with annual kilometers below a certain threshold. However, if motorists are below, or well above, this threshold they have no incentive to reduce driving.
[2] Government incentives may be needed to overcome obstacles to the private development of PAYD. When an insurer charges by the kilometer, its costs are reduced to the extent that its own customers reduce their accident risk by driving less. However, the costs to other insurance companies also are lowered because the risk of two-car accidents for their own customers is lower but savings cannot be captured by the company offering the kilometer-based insurance.

Policies to Address Obstacles to Clean Technologies

Even if corrective environmental taxes are feasible, most likely, owing to various obstacles that prevent sufficient investment in clean technologies, they are not sufficient. However, addressing technology barriers is largely tangential to the focus here on environmental tax design.

First, the most important policy, meaning the one yielding the biggest net benefits, usually is getting the prices right through corrective fiscal instruments, mainly because doing so provides across-the-board incentives for clean technology development and deployment. Further innovation incentives can yield significant, additional benefits, though studies suggest they are on a smaller scale (Goulder and Mathai, 2000; Nordhaus, 2002; Parry, Evans, and Oates, 2014; and Parry, Pizer, and Fischer, 2003).[23]

Second, because barriers vary in severity across different technologies, targeted measures are called for rather than setting environmental taxes in excess of environmental damage, which would encourage all technologies equally.[24]

The remainder of this subsection discusses the nature of technology barriers and possible responses, to complement environmental taxes, in the context of private sector research and development (R&D) and technology deployment.[25]

---

[23] A caveat is that delaying clean technology transitions is costly—and the costs grow if economies become even more locked into emissions-intensive capital and infrastructure (Acemoglu and others, 2012).
[24] Studies indicate that it is much less costly to promote emissions reductions and cleaner technologies by combining environmental taxes with technology incentives, rather than relying exclusively on taxes (Goulder and Schneider 1999; Fischer and Newell, 2008).
[25] Governments also conduct basic research into new technologies, the fruits of which are then used by the private sector. For example, the U.S. federal government spends about $4 billion a year on energy-related technologies, though a number of analysts believe that significantly more spending is warranted (Newell, 2008).

The focus is on altering private sector investment behavior rather than public investment (in transportation systems, fuel distribution infrastructure, smart grids, and the like). Generally, public investments should be warranted on cost-benefit criteria, taking into account their potential role in enhancing the effectiveness of environmental taxes, for example, by providing commuters with public transit alternatives (World Bank, 2012).

R&D

Private R&D into cleaner technologies is inadequate, even with corrective environmental taxes, when innovators are unable to capture the new technologies' benefits that spill over to other firms that might copy them or use them to further their own research programs. Uncertainty about future policy also makes firms hesitant to invest in new technologies. Although similar barriers might apply to technology development in other sectors, they seem especially severe for cleaner energy technologies (e.g., renewables plants) where upfront costs are often large and emissions reductions may persist for several decades; even with adequate corrective taxes now, tax rates in the far future are inherently uncertain.

One technology instrument is R&D subsidies, such as tax credits, though subsidies do not distinguish between more promising and not so promising research possibilities. Granting intellectual property rights is better in this respect, because the value of the patent depends on the commercial viability of the technology. But patents set up a tension between R&D incentives and diffusion—if it is easy for other firms to "imitate around" the patented technology, new technologies are more easily diffused, but returns to the original innovator are undermined. Prizes for new technologies may be a useful supplement because they avoid this tension. Awards for critical new technologies might be based on objective analysis (estimating, for example, how much the technology helps lower the costs of meeting climate objectives), or smaller rewards, based on potential emissions reductions, might be paid to the innovator each time the technology is adopted by another firm.

Technology deployment

Clean technologies may also be deployed insufficiently, despite emissions pricing and R&D incentives, for several reasons beyond future policy uncertainty. For example, individual firms may be reluctant to pioneer the use of an immature technology because they will incur the costs associated with learning how to use it reliably and efficiently, while benefits from this learning partially accrue to other firms that subsequently adopt the technology. And a variety of problems could arise at the household level, though the basis for policy intervention remains contentious (see Box 3.10).

BOX 3.10

The Energy Paradox Controversy

The "energy paradox" refers to the observation that seemingly cost-effective energy-saving technologies, whose lifetime fuel savings discounted at market rates exceed their upfront purchase and installation costs, are not always adopted in the marketplace.

Numerous explanations have been proposed for this phenomenon, many of which may justify policy action. For example, consumers may have limited information, limited ability to calculate future energy costs from the information they have, or may have more product characteristics to consider than they can process, and so omit energy savings. They may also be mistrustful of claimed energy cost savings, doubtful about future fuel prices, or short-sighted in their assessment of the future. Information gaps in second-hand product markets could perpetuate such short-sightedness by not allowing people to reap the full advantage of higher energy efficiency when selling used products. Consumers may also be subject to borrowing constraints causing them to underinvest in energy-saving technologies relative to what would be desirable from society's perspective.

Other explanations, however, do not warrant policy intervention. For example, the observed reluctance of consumers to pay for more-energy-efficient products may reflect their awareness of possible undesirable side effects, such as reduced acceleration for cars, inferior quality of lighting for fluorescent bulbs compared with incandescent, or greater likelihood of these products to need repairs.

However, evidence on the extent to which energy efficiency is undervalued, and if so, the extent of warranted policy intervention, remains inconclusive, making it difficult to draw solid recommendations about the appropriate role of additional policies to address the energy paradox (Allcott and Wozny, 2012; Busse, Knittel, and Zettelmeyer, 2012; Gillingham, Newell, and Palmer, 2009; Helfand and Wolverton, 2011; Huntington, 2011; Parry, Evans, and Oates, 2014; Sallee, 2013).

Although additional instruments to spur technology deployment are likely needed, their appropriate form, scale, and phasing in can be tricky to judge. Interventions might include supplementary measures such as feebates or regulations to improve vehicle fuel efficiency or to encourage the penetration of renewables or other technologies. In the latter case, adoption subsidies might be better than regulations that force in the technology regardless of economic conditions; consistent with the previous discussion, subsidies allow firms the flexibility to deploy the technology on a more limited basis should its costs turn out to be greater (relative to other options) than initially expected.

Overcoming Obstacles to Environmental Tax Reform

Implementing environmental tax reform is challenging, and one reason is opposition to higher energy prices. As noted in Chapter 2, rather than tax fossil fuel energy, many countries subsidize it, to the tune of $490 billion in 2011. Moreover, even in countries that tax energy heavily, the taxes are often relatively blunt from an environmental perspective. And, because of overlapping programs and lower rates for favored groups, different fuel users can be charged quite different rates for the same emissions sources.[26]

Opposition to higher energy prices comes from households (a particular concern is low-income households) and from energy-intensive firms, especially in trade-sensitive sectors. Each is discussed briefly below; the issues are extensively covered elsewhere (Clements and others, 2013; Dinan, forthcoming). Clements and others (2013) also use case studies to discuss broader possibilities for enhancing the prospects for energy price reform (e.g., improving transparency, phasing reforms, informational campaigns).

Compensating households

For some countries, one possibility for reducing household opposition to environmental tax reform is to scale back preexisting taxes affecting energy that are, at least on environmental grounds, made redundant by the environmental tax. For example, in most Organization for Economic Cooperation and Development countries, most if not all of the burden of carbon pricing on residential electricity consumers and motorists could be offset by reducing current excise taxes on electricity consumption and vehicle sales (IMF, 2011). Another possibility might be to provide transitory subsidies for the adoption of cleaner energy alternatives such as heat insulation, fluorescent lighting, and solar water heaters.

In advanced countries, poorer households tend to spend a relatively large portion of their income on electricity and fuels for transportation, heating, and cooking. Consequently, relative to their income the burden of higher energy prices tends to be greater for lower-income households than for wealthier households, which runs counter to broader efforts to moderate income inequality. For developing countries, the burden of higher energy prices (relative to income) might be smaller for lower-income groups, owing to limited vehicle ownership or grid access. But any policy that potentially reduces living standards for the poor may require some offsetting compensation.

Setting energy prices below levels warranted by production costs and environmental damage is usually a highly inefficient way to help the less well off. According to estimates summarized in Figure 3.3, only 3 percent and 7 percent, respectively, of the benefits from lower gasoline and diesel prices in countries that subsidize these fuels accrue to the bottom income quintile.

In other words, there are much more efficient (i.e., more targeted) ways to help these groups, such as the following:
- Targeted tax cuts like payroll tax rebates, earned income tax credits, and higher personal income tax thresholds in countries where large numbers of low-income, energy-dependent households are covered by such taxes

---

[26]In the United Kingdom, for example, Johnson, Leicester, and Levell (2010, Table 4.1), estimate that implicit CO_2 taxes in 2009 were £26/ton and £41/ton for natural gas used in domestic and business power generation, respectively, and £0/ton and £9/ton for natural gas used in homes and industry, respectively.

Figure 3.3 Distributional Incidence of Energy Subsidies

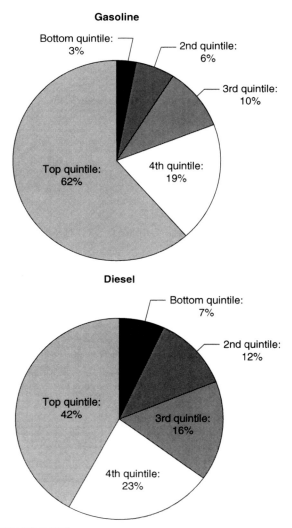

Source: Clements and others (2013).
Note: This figure summarizes, aggregated across all countries that subsidize gasoline and diesel fuels, the portion of benefits accruing to different income groups. Top quintile = richest; bottom quintile = poorest.

- Transfer payments, wage subsidies for low-paid jobs, and the like in countries where these compensation schemes are or could be administratively feasible
- Increased government spending (e.g., on schools, education, housing, jobs programs) that disproportionately benefits low-income households.

Recall, however, the caveats about overcompensation, and the preference for schemes that also promote economically desirable behavior.

Compensating firms

Higher energy costs for trade-exposed industries such as steel, aluminum, cement—a particular concern for carbon pricing—can lead to a dual problem: a loss of competitiveness, reflected in firms' relocating to countries where carbon pricing is not applied, and "emissions leakage" (increased emissions in these countries partly offsetting domestic reductions).[27] Possible responses, in the context of carbon pricing, include the following (Fischer, Morgenstern, and Richardson, 2013):

- Using some of the carbon pricing revenues to fund a general reduction in corporate income taxes provides offsetting gains in competitiveness for the economy as a whole, though these reductions are not well targeted to the most vulnerable energy-intensive firms and therefore do little to limit emissions leakage.
- Production subsidies can be used to protect energy-intensive, trade-sensitive firms by roughly offsetting their higher energy costs. These subsidies preserve incentives for reducing emissions per unit of output (though not for reducing their overall level of output). However, subsidies use up some of the carbon tax revenues and complicate administration.
- Border adjustments might be charged on embodied carbon in products imported to a country (or block of countries) with carbon pricing. A key attraction of these adjustments is that they encourage other trading countries to price carbon to avoid bearing some of the burden of tax accruals to other countries. The legality of such charges under free trade agreements is uncertain, however. Moreover, import fees may be tricky to implement (especially if applied to a large number of products, from many different countries) if embodied carbon and the extent to which other countries are mitigating carbon are difficult to measure.
- Yet another possibility is simply to exempt trade-sensitive firms—for example, by providing them a rebate for purchases of electricity and fuels to neutralize the effect of carbon pricing on their input costs. These exemptions also result in forgone revenue and, to a greater extent than for other measures, undermine environmental effectiveness.

In short, each of the above options has its drawbacks. Ideally, countries would coordinate their carbon pricing policies, lessening the pressure for such measures. (In any case, firms that cannot compete when domestic energy is appropriately priced should eventually be allowed to go out of business.)

---

[27]Studies (e.g., Böhringer, Carbone, and Rutherford, 2012) suggest that leakage offsets about 5–20 percent of the emissions reductions from carbon pricing, depending in part on the size of the coalition of countries taking action. This leakage reflects not only the international migration of economic activity, but also increases in fossil fuel use in other countries as world fuel prices fall in response to reduced demand in countries with carbon pricing. The latter type of leakage is not easily addressed through policy.

Applicability to Low-Income Countries

How applicable is the above discussion to low-income countries, where policymakers' primary concern may be to lift people out of abject poverty (rather than raise the costs of energy and transportation)?

With regard to climate change, low-income countries contribute very little to global emissions, and for practical purposes the case for them to undertake costly mitigation policies is correspondingly weak (Gillingham and Keen, 2012). But pricing for local environmental problems—air pollution, congestion, accidents—is in these countries' own interests because it provides net economic benefits. There are some nuances, however.

One is that the potential fiscal and environmental benefits are less important, in relative terms, for countries with relatively low energy intensity of GDP and that, as is common in Africa, do not use coal. Another is that, even with corrective taxes, and leaving aside the technology barriers already mentioned, private sector investment in green technologies may still be below economically efficient levels in low-income countries because of capital shortages. This is the basic rationale for donor contributions that support other investments such as infrastructure projects, and similar external funding has a complementary role to play in the environmental area. More generally, technology transfers to low-income countries can be promoted through dissemination of know-how acquired in advanced and emerging economies.

SUMMARY

Although this chapter provides a broad overview of instrument choice for environmental protection and policy design issues, for the purposes of the following chapters (which mainly focus on assessing efficient fuel tax levels) the main points are the following:

- For power generation, either (1) taxes should be levied on fuel supply (coal or natural gas) in proportion to emissions factors weighted by environmental damage per unit of emissions, with appropriate crediting for any emissions captured during fuel combustion or (2) charges should be directly levied on emissions released from smokestacks reflecting environmental damage per ton of emissions. The choice between these charging schemes largely hinges on administrative considerations.
- For heating fuels, charges should be levied on fuel supply to reflect emission rates and environmental damage (pricing emissions is not practical given the large number of fuel users).
- For transportation fuels, corrective taxes accounting for a wider range of side effects should be calculated according to equation (3.1).

REFERENCES

Acemoglu, Daron, Philippe Aghion, Leonardo Bursztyn, and David Hemous, 2012, "The Environment and Directed Technological Change," *American Economic Review*, Vol. 102, pp. 131–66.

Allcott, Hunt, and Nathan Wozny, 2013, "Gasoline Prices, Fuel Economy, and the Energy Paradox," NBER Working Paper No. 18583 (Cambridge, Massachusetts: National Bureau of Economic Research).

Böhringer, C., J.C. Carbone, and T.F. Rutherford, 2012, "Unilateral Climate Policy Design: Efficiency and Equity Implications of Alternative Instruments to Reduce Carbon Leakage," *Energy Economics*, Vol. 34 (Supplement 2), pp. S208–S217.

Busse, Meghan R., Christopher R. Knittel, and Florian Zettelmeyer, 2012, "Are Consumers Myopic? Evidence from New and Used Car Purchases," *American Economic Review*, Vol. 103, pp. 220–56.

Cairns, Robert D., 2014, "The Green Paradox of Exhaustible Resources," *Energy Policy*, Vol. 65, February, pp. 78–85.

CE Delft, Infras, Faaunhofer ISI, 2011, *External Costs of Transport in Europe: Update Study for 2008* (Delft, Netherlands: CE Delft).

Clements, Benedict, David Coady, Stefania Fabrizio, Sanjeev Gupta, Trevor Alleyene, and Carlo Sdralevich, eds., 2013, *Energy Subsidy Reform: Lessons and Implications* (Washington: International Monetary Fund).

de Borger, Bruno, and Stef Proost, 2001, *Reforming Transport Pricing in the European Union* (Northampton, Massachusetts: Edward Elgar).

Delucchi, Mark A., 2000, "Environmental Externalities of Motor Vehicle Use," *Journal of Transport Economics and Policy*, Vol. 34, pp. 135–68.

Department of Climate Change and Energy Efficiency, 2011, *Securing a Clean Energy Future: The Australian Government's Climate Change Plan* (Canberra, Australia: Department of Climate Change and Energy Efficiency).

Dinan, Terry, forthcoming, "Offsetting a Carbon Tax's Burden on Low-Income Households," in *Implementing a U.S. Carbon Tax: Challenges and Debates*, edited by I. Parry, A. Morris, and R. Williams (Washington: International Monetary Fund).

Eliasson, Jonas, 2009, "A Cost–Benefit Analysis of the Stockholm Congestion Charging System," *Transportation Research Part A: Policy and Practice*, Vol. 43, pp. 468–80.

Evans, Carol A., Clifford Winston, and Kenneth A. Small, 1989, *Road Work: A New Highway Pricing and Investment Policy* (Washington: Brookings Institution).

Fischer, Carolyn, Richard Morgenstern, and Nathan Richardson, 2013, "Carbon Taxes and Energy Intensive Trade Exposed Industries: Impacts and Options," Working paper (Washington: Resources for the Future).

Fischer, Carolyn, and Richard G. Newell, 2008, "Environmental and Technology Policies for Climate Mitigation," *Journal of Environmental Economics and Management*, Vol. 55, pp. 142–62.

Fischer, Carolyn, Ian W.H. Parry, and Winston Harrington, 2007, "Should Corporate Average Fuel Economy (CAFE) Standards be Tightened?" *Energy Journal*, Vol. 28, pp. 1–29.

Gillingham, Kenneth, Richard G. Newell, and Karen Palmer, 2009, "Energy Efficiency Economics and Policy," Discussion Paper No. 09–13 (Washington: Resources for the Future).

Gillingham, Robert, and Michael Keen, 2012, "Mitigation and Fuel Pricing in Developing Countries," in *Fiscal Policy to Mitigate Climate Change: A Guide for Policymakers*, edited by I.W.H. Parry, R. de Mooij, and M. Keen (Washington: International Monetary Fund).

Goulder, Lawrence H., ed., 2002, *Environmental Policy Making in Economies with Prior Tax Distortions* (Northampton, Massachusetts: Edward Elgar).

———, and Koshy Mathai, 2000, "Optimal CO_2 Abatement in the Presence of Induced Technological Change," *Journal of Environmental Economics and Management*, Vol. 39, pp. 1–38.

Goulder, Lawrence H., and Ian W.H. Parry, 2008, "Instrument Choice in Environmental Policy," *Review of Environmental Economics and Policy*, Vol. 2, pp. 152–74.

———, Roberton C. Williams, and Dallas Burtraw, 1999, "The Cost-Effectiveness of Alternative Instruments for Environmental Protection in a Second-Best Setting," *Journal of Public Economics*, Vol. 72, pp. 329–60.

Goulder, Lawrence H., and Stephen H. Schneider, 1999, "Induced Technological Change and the Attractiveness of CO_2 Abatement Policies," *Resource and Energy Economics*, Vol. 21, pp. 211–53.

Government of British Columbia, 2012, *Budget and Fiscal Plan 2012/13–2014/15* (Vancouver, Canada: Ministry of Finance, Government of British Columbia).

Helfand, Gloria, and Ann Wolverton, 2011, "Evaluating the Consumer Response to Fuel Economy: A Review of the Literature," *International Review of Environmental and Resource Economics*, Vol. 5, pp. 103–46.

Hepburn, Cameron, 2006, "Regulation by Prices, Quantities, or Both: A Review of Instrument Choice," *Oxford Review of Economic Policy*, Vol. 22, pp. 226–47.

Huntington, Hillard, 2011, "The Policy Implications of Energy-Efficiency Cost Curves," *Energy Journal*, Vol. 32, pp. 7–22.

International Monetary Fund (IMF), 2008, "The Fiscal Implications of Climate Change" (Washington: International Monetary Fund).

———, 2011, "Promising Domestic Fiscal Instruments for Climate Finance," Background Paper for the Report to the G20 on *Mobilizing Sources of Climate Finance* (Washington: International Monetary Fund).

Johnson, Paul, Andrew Leicester, and George Stoye, 2012, *Fuel for Thought: The What, Why and How of Motoring Taxation* (London: Institute for Fiscal Studies).

Jones, Ben, and Michael Keen, 2011, "Climate Policy in Crisis and Recovery," *Journal of International Commerce, Economics and Policy*, Vol. 2, pp. 325–49.

Krupnick, Alan J., Ian W.H. Parry, Margaret Walls, Tony Knowles, and Kristin Hayes, 2010, *Toward a New National Energy Policy: Assessing the Options* (Washington: Resources for the Future and National Energy Policy Institute).

Maibach, M., C. Schreyer, D. Sutter, H.P. van Essen, B.H. Boon, R. Smokers, A. Schroten, and C. Doll, 2008, *Handbook on Estimation of External Costs in the Transport Sector* (Delft, The Netherlands: CE Delft).

Muller, Nicholas Z., and Robert Mendelsohn, 2009, "Efficient Pollution Regulation: Getting the Prices Right," *American Economic Review*, Vol. 99, pp. 1714–39.

Newell, Richard G., 2008, "A U.S. Innovation Strategy for Climate Change Mitigation" (Washington: The Hamilton Project, Brookings Institution).

———, and Robert N. Stavins, 2003, "Cost Heterogeneity and the Potential Savings for Market-Based Policies," *Journal of Regulatory Economics*, Vol. 23, pp. 43–59.

Nordhaus, William D., 2002, "Modeling Induced Innovation in Climate-Change Policy," in *Technological Change and the Environment*, edited by Arnulf Grubler, Nebojsa Nakicenovic, and William Nordhaus (Washington: Resources for the Future) pp. 182–209.

Oates, Wallace E., and Diana L. Strassmann, 1984, "Effluent Fees and Market Structure," *Journal of Public Economics*, Vol. 24, pp. 29–46.

Opschoor, J.B., and H.B. Vos, 1989, *Economic Instruments for Environmental Protection* (Paris: Organization for Economic Cooperation and Development).

Organization for Economic Cooperation and Development (OECD), 2010, *Taxation, Innovation and the Environment* (Paris: Organization for Economic Cooperation and Development).

Parry, Ian W.H., David Evans, and Wallace E. Oates, 2014, "Are Energy Efficiency Standards Justified?" *Journal of Environmental Economics and Management*, Vol. 67, pp. 104–25.

Parry, Ian W.H., and Wallace E. Oates, 2000, "Policy Analysis in the Presence of Distorting Taxes," *Journal of Policy Analysis and Management*, Vol. 19, No. 4, pp. 603–14.

Parry, Ian W.H., William A. Pizer, and Carolyn Fischer, 2003, "How Large Are the Welfare Gains from Technological Innovation Induced by Environmental Policies?" *Journal of Regulatory Economics*, Vol. 23, pp. 237–55.

Parry, Ian W.H., and Kenneth A. Small, 2005, "Does Britain or the United States Have the Right Gasoline Tax?" *American Economic Review*, Vol. 95, No. 4, pp. 1276–89.

Parry, Ian W.H., and Roberton C. Williams, 2012, "Moving US Climate Policy Forward: Are Carbon Tax Shifts the Only Good Alternative?" in *Climate Change and Common Sense: Essays in Honor of Tom Schelling*, edited by Robert Hahn and Alistair Ulph (Oxford, U.K.: Oxford University Press) pp. 173–202.

Prust, Jim, and Dominique Simard, 2004, "U.S. Energy Policy: The Role of Taxation," in *U.S. Fiscal Policies and Priorities for Long-Run Sustainability*, edited by Martin Mühleisen and Christopher Towe (Washington: International Monetary Fund).

Quinet, Emile, 2004, "A Meta-Analysis of Western European External Costs Estimates," *Transportation Research Part D*, Vol. 9, pp. 465–76.

Sallee, James M., 2013, "Rational Inattention and Energy Efficiency," NBER Working Paper No. 19545 (Cambridge, Massachusetts: National Bureau of Economic Research).

———, Hannah Behrendt, Laura Maconi, Tara Shirvani, and Alexander Teytelboym, 2010, "Part I: Externalities and Economic Policies in Road Transport," *Research in Transportation Economics*, Vol. 28, pp. 2–45.

Santos, Georgina, 2005, "Urban Congestion Charging: A Comparison between London and Singapore," *Transport Reviews*, Vol. 25, pp. 511–34.

Sinn, Hans-Werner, 2012, *The Green Paradox: A Supply-Side Approach to Global Warming* (Cambridge, Massachusetts: MIT Press).

Small, Kenneth A., 2010, "Energy Policies for Automobile Transportation: A Comparison Using the National Energy Modeling System" (Washington: Resources for the Future).

———, and Kurt Van Dender, 2006, "Fuel Efficiency and Motor Vehicle Travel: The Declining Rebound Effect," *Energy Journal*, Vol. 28, No. 1, pp. 25–52.

T&E, 2011, *Fuelling Oil Demand: What Happened to Fuel Taxation in Europe?* (Brussels: European Federation for Transport and Environment).

Transport for London, 2008, *Central London Congestion Charging: Impacts Monitoring*, Sixth Annual Report (London: Transport for London).

United States Federal Highway Administration (US FHWA), 1997, *1997 Federal Highway Cost Allocation Study* (Washington: Federal Highway Administration, US Department of Transportation).

Weitzman, Martin L., 1974, "Prices vs. Quantities," *Review of Economic Studies*, Vol. 41, pp. 477–91.

World Bank, 2011, *Air Quality Analysis of Ulaanbaatar: Improving Air Quality to Reduce Health Impacts* (Washington: World Bank).

———, 2012, *Inclusive Green Growth: The Pathway to Sustainable Development* (Washington: World Bank).

CHAPTER 4

Measuring Pollution Damage from Fuel Use

This chapter begins with a brief review of the literature on valuing climate change damage from carbon dioxide (CO_2) emissions. The heart of the chapter is about measuring damage from the most important harm from local air pollution: human mortality risk.

CO_2 DAMAGE

The future climate change damage from a ton of CO_2 emissions is the same regardless of the fuel combustion process or where emissions are released. In principle, therefore, each ton should be priced the same in different countries. If charges are imposed on fuel suppliers, the appropriate charge per unit of fuel is CO_2 damage times the CO_2 emissions factor (i.e., CO_2 emissions released per unit of fuel combustion). The first component is discussed here, and emissions factors are discussed in the second section.

Two economic approaches have been used to assess appropriate CO_2 emissions prices—the benefit-cost and cost-effectiveness approaches.[1]

Benefit-Cost Approach

The benefit-cost approach assesses damage from future global climate change caused by additional emissions using "integrated assessment models" that incorporate the following:

- Links between current emissions and the future global time-path of atmospheric greenhouse gas (GHG) concentrations
- Impacts of changes in that time-path on global temperature and other climate variables in future years
- Worldwide monetized damage from those climate changes (e.g., agricultural impacts, costs of sea-level protection, health impacts from altered climate and possible spread of vector-borne diseases, ecological impacts)
- Discounting of damage at different future dates to the present, to obtain a single summary statistic, or damage per ton of CO_2, known as the "social cost of carbon" (SCC).

---

[1]For more extensive discussions see, for example, Bosetti and others (2012); Griffiths and others (2012); National Research Council (2009), Ch. 5; and US IAWG (2013).

Although many uncertainties surround these relationships, damage values are especially sensitive to discounting (CO_2 emissions have very long range impacts because they reside in the atmosphere for many decades and the climate adjusts gradually to higher atmospheric concentrations) and the treatment of extreme risks.

One view on discounting is that the future benefits of emissions mitigation policies should be discounted using market interest rates (usually about 3–5 percent in advanced countries) because this is the standard way to evaluate the future benefits of any private and many public investments. Studies using market discount rates typically estimate the SCC to be about $10/ton to $50/ton of CO_2; for example, US IAWG (2013) puts the SCC at $35/ton (in their central case for 2010 in 2010 dollars).

Others argue that for ethical reasons, below market rates should be used to evaluate policies if the benefits accrue to future generations (as opposed to the current generation), to avoid discriminating against people who are not yet born. Under this approach, SCC estimates are much higher, for example, Stern (2007) puts the SCC at $85 per ton (in 2004 dollars), with the difference compared with earlier studies largely reflecting different discount rates (Nordhaus, 2007).[2]

These SCC estimates often include a component for catastrophic risks by postulating probabilities—based on judgment, given that the risks are unknown—that future climate change may result in very large world GDP losses. However, the appropriate way to treat these risks remains very contentious (Pindyck, 2013): some studies (e.g., Weitzman, 2009) suggest they warrant dramatically higher CO_2 prices.[3]

Typically, SCC estimates in the benefit-cost approach increase about 1.5–2.5 percent a year in real terms, primarily reflecting the growth rate in output potentially affected by climate change.

Cost-Effectiveness Approach

Rather than explicitly valuing environmental damage (the approach taken elsewhere in this volume), the cost-effectiveness approach assesses least-cost pricing paths for CO_2 emissions that are broadly consistent with long-term climate stabilization goals.

Numerous climate-economy models have been developed, with particular detail on the global energy sector and links between emissions, atmospheric GHG concentrations, and future climate outcomes. Projecting future emissions prices needed to meet long-range climate targets is inherently imprecise, however, given considerable uncertainty about future emissions baselines (which depend on future population, per capita income, the energy-intensity of GDP, the fuel mix, and so on) and the emissions impact of pricing (which depends on the future costs of low-emission fuels and technologies and other factors).

---

[2] The discount rate in this approach is still positive (usually about 1–2 percent) to reflect the higher per capita consumption of future generations.
[3] Damage estimates are highly sensitive to the shape of the probability distribution for catastrophic risks, which is uncertain.

A global CO_2 price starting at about $30/ton (in current dollars) in 2020 and rising about 5 percent a year would be roughly in line with ultimately containing mean projected warming to 2.5°C (Nordhaus, 2013, p. 228) at least cost. A substantially higher global price would be needed to contain mean projected warming to 2°C (the goal identified in the 2009 Copenhagen Accord), though this target might now be beyond reach because it would likely require the future development and global deployment of technologies that, on net, remove GHGs from the atmosphere, to help lower future GHG concentrations to current levels. For any given climate stabilization target, substantially higher starting prices will also be needed in the absence of full participation by all the major emitting countries.

Illustrative Value Used Here

Tremendous uncertainty and controversy surround the appropriate CO_2 emissions price—and country governments may have their own perspectives. A value of $35/ton (based on US IAWG, 2013) is used here *for illustrative purposes.* This chosen value should not be construed as a recommendation for one SCC value over another or one climate stabilization target over another. The implications of alternative values for corrective taxes are easily inferred by proportionally scaling the component for carbon damages.

A Note on Equity

A key principle in the UN Framework Convention on Climate Change is that developing countries have "common but differentiated responsibilities," meaning (given their relatively low income and small contribution to historical GHG accumulations) that they should bear a disproportionately lower burden of mitigation costs than wealthier nations. This principle implies either their receiving compensation or their imposing lower emissions prices than others, or perhaps no price at all. Application of this principle need not hinder international mitigation efforts however, at least for the vast majority of low-income countries whose emissions constitute a tiny fraction of the global total (Gillingham and Keen, 2012).

LOCAL AIR POLLUTION DAMAGE

Although local air pollution causes a variety of other harmful environmental effects, the central issue is premature human mortality, which is, by far, the most important category in previous damage assessments (Chapter 2).

The pollution-mortality impacts from fuel combustion can be valued using the following steps:
- Determining how much pollution is inhaled by exposed populations, both in the country where emissions are released and, for emissions released from tall smokestacks, in countries to which pollution may be transported
- Assessing how this pollution exposure affects mortality risks, accounting for factors, such as the age and health of the population, that affect vulnerability to pollution-related illness

- Monetizing the health effects
- Expressing the resulting damage per unit of fuels.

The focus here is damage from an incremental amount of pollution rather than damage from the total amount of pollution because the incremental amount is relevant for setting efficient fuel taxes.

For a very limited number of countries, previous studies have estimated local air pollution damage, and major modeling efforts are ongoing at the global level.[4] This volume is the first attempt to provide an assessment of fossil fuel emissions damage across a broad range of developed and developing countries, using a consistent methodology.[5]

Although insofar as possible key country-specific factors determining environmental damage is captured, not all potentially significant factors, most notably cross-country differences in meteorological conditions affecting pollution formation, can feasibly be included. The corrective tax estimates in this chapter may also become outdated as evidence and data evolve. Nonetheless, some broad sense of how missing factors may affect the results is given by comparing the results for selected countries to those from a computational model of regional air quality. And accompanying spreadsheets,[6] indicating corrective taxes by fuel product and country, are easy to update.

The discussion proceeds as follows: The first three subsections address, respectively, the first three steps in the bulleted list above. The fourth subsection summarizes the resulting cross-country estimates of local pollution damage. The fifth subsection compares the results with those from the computational model. The final subsection discusses procedures for converting emissions damage into corrective fuel taxes, the results of which are presented in Chapter 6.

Estimating Population Exposure to Pollution

As noted in Chapter 2, the main cause of mortality risk from pollution is particulate matter with diameter up to 2.5 micrometers ($PM_{2.5}$), which is small enough to permeate the lungs and bloodstream. $PM_{2.5}$ can be emitted directly as a primary pollutant from fuel combustion, but is also produced as a secondary pollutant from chemical reactions in the atmosphere involving primary pollutants, the most important of which is sulfur dioxide (SO_2), but also nitrogen oxides (NO_x).

---

[4]For example, National Research Council (2009) and Muller and Mendelsohn (2012) estimate pollution damage for the United States; European Commission (2008) for Europe; World Bank and State Environmental Protection Agency of China (2007) for China; Cropper and others (2012) for India. At the global level, the Global Burden of Disease project (discussed later in this chapter) estimates regional mortality rates from air pollution, though not the health effects from emissions released in individual countries, which is needed to derive corrective fuel taxes. The Climate and Clean Air Coalition is developing a sophisticated modeling system to quantify air pollution damage. The model presently covers four countries but will eventually apply to many more. As this work progresses, it will provide useful information for refining corrective tax estimates.

[5]Methodological consistency implies that differences in estimated damage across countries are more likely to reflect real factors rather than different estimation procedures.

[6]Available at www.imf.org/environment.

"Intake fractions" are used to estimate how much pollution from stationary and mobile emissions sources in different countries is inhaled by exposed populations (see Box 4.1 for technical details). Specifically, these fractions, as used here, indicate grams of $PM_{2.5}$ inhaled per ton of primary $PM_{2.5}$, SO_2, and NO_x. Intake fractions are a powerful concept and are being used increasingly in pollution damage assessment (Apte and others, 2012; Bennett and others, 2002; Cropper and others, 2012; Humbert and others, 2011; Levy, Wolff, and Evans, 2002; Zhou and others, 2006), primarily because they circumvent the need to develop data- and computationally intensive air quality models.

Intake fractions depend on three main factors:

- *The height at which emissions are released:* The most important distinction is between emissions from tall smokestacks, such as at power plants, which are more likely to be dispersed without harm but are also transported considerable distances, and emissions released at ground level, such as from cars and residential heating, which tend to stay locally concentrated.

- *The size of the population exposed to the pollution:* For smokestack emissions, people living 2,000 kilometers or more from a plant can still intake some of the pollution (Zhou and others, 2006). Even if a plant were to be located away from an urban center, its emissions could still cause significant health damage elsewhere. Long-distance transportation of pollution also raises thorny issues about how one country should account for cross-border environmental damage when setting its own fuel taxes.

BOX 4.1

Intake Fractions: Some Technicalities

The intake fraction (iF) for a primary pollutant at a particular location is given by the formula (Levy, Wolff, and Evans, 2002):

$$iF \equiv \frac{\sum_{i=1}^{N} P_i \times \Delta C_i \times BR}{Q},$$

in which P_i is the population residing in a region, indexed by i, defined by its distance from the emissions source; the region could be in the country where emissions are released or in some other country or some combination of both. The term ΔC_i is the change in the ambient concentration of pollution ($PM_{2.5}$), perhaps defined by the daily average change in pollution per cubic meter, caused by emissions from the source; ΔC_i is influenced by meteorology and other factors. BR is the average breathing rate, that is, the rate at which a given amount of ambient pollution is inhaled by the average person; for example, in Zhou and others (2006) the breathing rate is 20 cubic meters per day.

The numerator in the equation is, therefore, the total daily amount of pollution taken in by potentially exposed populations. In the denominator, Q is the emission rate of the primary pollutant in tons per day. The intake fraction is defined as average pollution inhaled per unit of emissions released, and is usually expressed as grams of $PM_{2.5}$ inhaled per ton of primary emissions.

- *Meteorological conditions* (most notably wind speed and direction), *topography* (e.g., proximity to mountain barriers that may block pollution dispersion), and *ambient ammonia concentrations* (which catalyze atmospheric reactions of SO_2 and NO_x).

For long-distance pollution, a strength of the approach used here is that it uses highly disaggregated data on population density (in different countries) up to 2,000 kilometers away from emissions sources. Therefore, the estimates of population exposure may be considerably more accurate than in other studies using much more spatially aggregated population data, or that only consider people living within shorter distances of the emissions source.

A weakness is that the intake fraction approach cannot easily account for cross-country differences in meteorological and related conditions, not least because emissions are transported across multiple climate zones and wind patterns. However, studies suggest that population exposure is usually, by far, the more important factor (Zhou and others, 2006).

The estimation of population exposure for coal plants, other stationary sources, and mobile emissions sources are discussed in turn below.

Exposure to coal plant emissions

Although intake fractions have been extensively estimated for emissions released at ground level for many different regions (see the discussion of mobile sources below), estimates are much more limited for emissions released from tall smokestacks because of the complexities involved in modeling long-distance pollution transport.

The approach in this chapter uses a widely cited study by Zhou and others (2006), which follows a two-step statistical procedure. Using a sophisticated model of regional air quality, they start by simulating how emissions are transported to different regions, then map the result to data on regional population density, to estimate intake fractions for a variety of primary pollutants from 29 coal plants in China.[7] For example, for the average coal plant, they estimate that about 5 grams of $PM_{2.5}$ ends up being inhaled for each ton of SO_2 emitted. Zhou and others (2006) then use statistical techniques to obtain a set of coefficients indicating what fraction of an average plant's emissions are inhaled by an average person residing within bands of 0–100 kilometers, 100–500 kilometers, 500–1,000 kilometers, and 1,000–3,300 kilometers from the emissions source.

These coefficients can be combined with data on the number of people living within the four distance classifications from the plant to extrapolate intake fractions for a coal plant in any country, without the need for developing a sophisticated model of regional air quality. To keep the calculations tractable, the last

---

[7] Zhou and others (2006) use the California Puff (CALPUFF) air quality model, calibrated to Chinese data on regional emissions sources and pollution concentrations. This model is recommended by the US Environmental Protection Agency for estimating long-distance pollution transport (for documentation, see www.src.com/calpuff/calpuff1.htm).

distance category is truncated for the analysis in this chapter (without much loss of accuracy) at 2,000 kilometers.[8]

For extrapolation purposes, the Carbon Monitoring for Action (CARMA) database[9] is used to determine the geographical location of about 2,400 coal plants in about 110 different countries for 2009 (these data cover about 75 percent of the total electricity produced by coal power plants worldwide).

LandScan data are used to obtain 2010 population counts by grid cell for each of these 110 countries, as well as for countries without coal plants but where people are still at risk of inhaling cross-border emissions.[10] These population data are extremely fine—each grid cell is 1 kilometer square or less.

Mapping these two data sets provides an extremely accurate estimate of the population living at the four distance classifications from each plant. Multiplying populations in these distance categories by the corresponding coefficient from Zhou and others (2006) for a particular pollutant, and then adding over the four distance categories, gives the estimated intake fraction for that pollutant for each coal plant.

Finally, the national average intake fraction for the pollutant is obtained for each country by taking a weighted sum of intake fractions for individual plants in that country, where the weights are each plant's share in total coal use.[11]

There are some caveats to the intake fraction approach as applied to long-distance (but not ground-level) pollution. Most notably, adjustments are not made for meteorological or for topographical conditions or local ammonia concentrations. The last factor is relevant because SO_2 and NO_x form $PM_{2.5}$ by reacting with ammonia—in fact, when SO_2 and NO_x are reduced substantially, ammonia is "freed up" for the remaining SO_2 and NO_x emissions to react with, making them more likely to form fine particulates ($PM_{2.5}$). To the extent that all these factors vary (across a radius of 2,000 kilometers) for the average coal plant in another country relative to these conditions for the average coal plant in China, these estimates overstate or understate intake fractions for other countries. This issue is discussed further in a later section that reviews results for selected countries and compares them with those from a computational air quality model that does account for meteorological conditions, ammonia concentrations, and related factors.

---

[8] Zhou and others (2006) estimate that the inhalation rate (for SO_2-induced $PM_{2.5}$) for someone living within 100 kilometers of a coal plant is about 8 times that for someone living 100–500 kilometers away, 43 times that for someone living 500–1,000 kilometers away, and 86 times that for someone living 1,000–3,300 kilometers away. However, taking into account the average number of people living at different distance classifications, they find that people living within 100 kilometers, 100–500 kilometers, 500–1,000 kilometers and 1,000–3,300 kilometers of the plant inhale 53, 27, 6, and 14 percent, respectively, of the total pollution intake.
[9] See http://carma.org.
[10] The LandScan data are compiled by Oak Ridge National Laboratory (see www.ornl.gov/sci/landscan).
[11] Total coal use is the sum of coal used by plants in the sample for that country. Coal use is derived from the plant's CO_2 emissions given that there is a proportional relationship between fuel use and CO_2 emissions, and emissions at the plant level are reported in the CARMA database.

Second, intake fractions may also depend on the precise height of the smokestack from which pollutants are emitted, with emissions from the tallest smokestacks having the greatest propensity to dissipate before they are inhaled. Again, data are not available on global variation in the height of smokestacks at power plants to allow an adjustment to be made. However, intake fractions do not appear to vary much with differences in smokestack height (Humbert and others, 2011).

Third, mortality risks to people living close to two or more power plants are assumed to be additive (or in other words, the intake fraction for one coal plant is the same, regardless of whether some of the people inhaling its pollution are also exposed to pollution from other plants). For the most part this seems reasonable, except, perhaps, for countries where air pollution is especially severe and people's ability to inhale pollution starts to become saturated, but even then (see Box 3.3) there may not be much relevance for corrective fuel tax estimates.

Exposure to other stationary source emissions and vehicle emissions

Because of the lack of data—particularly in regard to geographical location—it is not feasible to estimate population exposure to emissions from other uses of coal (e.g., metals smelting). For purposes of calculating the impact of coal tax reform, environmental damage and corrective taxes for these other uses are assumed to be the same as for power plant coal use.

Essentially the same procedure and data sources as outlined above are used to estimate average population exposure to approximately 2,000 natural gas plants in 101 countries. Natural gas produces the same three primary pollutants as coal so the Zhou and others' (2006) coefficients can be applied again.[12]

Intake fractions for each primary pollutant tend to be greater for natural gas than for coal because, on average, gas plants are located closer to population centers, but the differences are not large. In fact, the local pollution effects of natural gas combustion are far less severe than for coal because natural gas produces minimal amounts of SO_2 and primary $PM_{2.5}$.

With respect to natural gas use in homes, primarily for space heating, population exposure to outdoor pollution is far more localized, given that emissions are released and stay close to the ground. The same applies for vehicle emissions.

For both residential and vehicle emissions, estimates from Humbert and others (2011) and Apte and others (2012) are combined. Humbert and others (2011) report a global average intake fraction for ground-level sources of SO_2, NO_x, and primary $PM_{2.5}$. Apte and others (2012) estimate, but only for primary $PM_{2.5}$, intake fractions for 3,646 urban centers across the world, accounting for local population density and meteorology.[13] The city-level intake fractions for

---

[12]Certain characteristics of natural gas plants, such as the rate and temperature at which emissions are released, may differ somewhat from those for coal but, most likely, this causes little bias in the intake fraction estimates for natural gas emissions.

[13]Although Apte and others (2012) mainly focus on vehicle emissions, their results apply broadly to any ground-level source of these emissions.

primary $PM_{2.5}$ (or a simple average of them for countries with more than one city in the data of Apte and others, 2012) are extrapolated to the country level by weighting them by the fraction of the population living in the relevant urban area.[14] Intake fractions for SO_2 and NO_x by country are then derived from Humbert and others' (2011) estimates, scaling them by the ratio of the intake fraction for $PM_{2.5}$ for that country to the global average intake fraction for $PM_{2.5}$ from Apte and others (2012).[15]

From Pollution Exposure to Mortality Risk

This section discusses the two steps needed to assess how additional pollution exposure increases mortality risk in different countries. The first step is to establish the baseline mortality rate for illnesses potentially aggravated by pollution. The second is to multiply these baseline mortality rates by estimates of the increased likelihood of mortality with extra pollution relative to mortality without extra pollution, and then aggregate over illnesses.

Much of the discussion relies on work by the World Health Organization's Global Burden of Disease project, which provides the most comprehensive assessment to date of mortality and loss of health from pollution-related and other diseases, injuries, and risk factors for all regions of the world.[16]

Baseline mortality rates

The increased mortality risk from extra pollution inhaled by a population of given size will depend on the age and health of the population. Seniors, for example, are generally more susceptible to pollution-induced illnesses than younger adults. Health status also matters—someone already suffering from a heart or lung condition that is potentially aggravated by inhaling pollution is more vulnerable than a healthy person. And if people are more likely to die prematurely from other causes (e.g., traffic accidents, non-pollution-related illness), they are, by definition, less likely to live long enough to die from pollution-related illness.

The role of these factors can be summarized by calculating an age-weighted mortality rate for illnesses potentially worsened by pollution. The focus is on the four adult diseases—lung cancer, chronic obstructive pulmonary disease, ischemic heart disease (from reduced blood supply), and stroke—all of whose prevalence is increased when people intake pollution.

---

[14]The urbanization rate by country for 2010 was obtained from World Bank (2013). Intake fractions for ground-level emissions in rural areas are not available on a country-level basis, though they will be considerably smaller than those for urban areas. Although it makes little difference for the results, an estimate of rural intake fractions is included based on a single global-level estimate reported in Humbert and others (2011). This estimated intake fraction is then weighted by the rural population share for each country and added to the intake fraction estimates for urban areas.

[15]Motor vehicles also produce volatile organic compounds (VOCs), the primary effect of which is to form ozone. Ozone damage is ignored here, however, given the much weaker link between ozone and mortality compared with $PM_{2.5}$.

[16]See www.who.int/healthinfo/global_burden_disease/about/en.

Annual mortality rates from these four illnesses were estimated for each country, taking into account the age structure of the population, as follows: Global Burden of Disease data provide mortality rates for the four diseases for 12 different age classifications at the regional level, with the world divided into 21 regions (the Annex Table 4.1.1 lists the countries within each region). These age classes are for people 25 and older (mortality risks for those younger than 25 are assumed to be zero—see below). Age-weighted mortality rates by disease at the country level are then obtained using the share of the country's population in each age class.[17]

Figure 4.1 shows the results for the 21 regions. At a global level, the total mortality rate for diseases potentially worsened by pollution is 3.7 deaths per 1,000 people per year. (Most of these deaths, roughly 89 percent on average, would still occur with no pollution.) Eastern Europe has the highest mortality rate, 10.6 deaths per 1,000 people, in part because of the high prevalence of alcohol- and smoking-related illness. The lowest mortality rate is 1.3 deaths per 1,000 people in western sub-Saharan Africa, where people are more prone to die from other causes rather than surviving long enough to suffer pollution-related illness.[18]

For all regions, heart disease is the largest source of mortality—at a global level it accounts for almost half of total deaths from the four diseases, with pulmonary disease and stroke accounting for about 20 percent each, and lung cancer about 10 percent. These shares vary somewhat by region—in Eastern Europe, for example, heart disease accounts for 72 percent of total deaths.

Pollution damage estimated in this volume is understated in the sense that premature deaths of those younger than 25, most notably from infant mortality, are excluded. One reason for omitting these deaths is that the valuation of mortality risk for infants is even more unsettled and contentious than that for adults (see Box 4.3).[19]

Increased mortality from air pollution

A limited number of studies for the United States have estimated the relationship between pollution concentrations and increased mortality for pollution-related diseases—so-called concentration response functions.[20] For example, Pope and

---

[17] The mortality data were obtained from http://ghdx.healthmetricsandevaluation.org/global-burden-disease-study-2010-gbd-2010-data-downloads. The population share data, by country, needed for these calculations are from http://unstats.un.org/unsd/demographic/products/dyb/dyb2.htm.

[18] The baseline rates for pollution-related mortality are based on studies estimating health impacts from all sources of air pollution, including biomass and natural sources (e.g., wind-blown dust and sea salt) in addition to fossil fuel emissions. However, generally speaking, other sources appear to be of relatively modest importance (Ostro, 2004; Schaap and others, 2010; Lükewille and Viana, 2012).

[19] Given that only pollution inhaled by people ages 25 and older potentially has health effects in the assessment, the intake fractions for each country are first multiplied by the share of people ages 25 and older in the total population for that country, before applying the pollution-health relationships discussed later.

[20] To apply these relationships, intake fractions are first divided by the breathing rate to convert them from pollution inhaled to pollution exposure (see Box 4.1).

Figure 4.1 Baseline Mortality Rates for Illnesses Whose Prevalence Is Aggravated by Pollution, Selected Regions, 2010

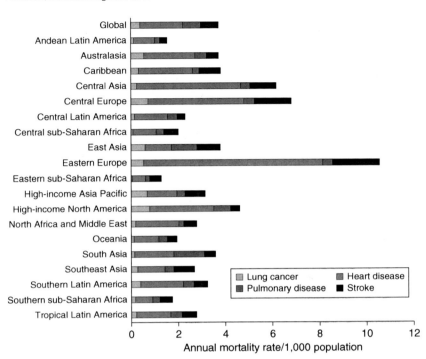

Source: Authors' calculations.
Note: Figure shows number of people (ages 25 and older) per 1,000 population dying from diseases whose incidence can be increased by outdoor air pollution. Only a minor portion should be attributed to pollution—the rest would occur anyway, even if there were no pollution.

others (2002) track the health status of a large cohort of adults in 61 U.S. cities over a long period to attribute health outcomes to $PM_{2.5}$ concentrations as opposed to other factors such as age, gender, income, dietary habits, smoking prevalence. They estimate that each 10 microgram/cubic meter increase in $PM_{2.5}$ concentrations increases annual mortality risks from all pollution-related illness in the United States by 6.0 percent. Until recently, the concentration-response functions underlying Pope and others (2002) were used in regulatory assessments by the United States Environmental Protection Agency (US EPA). However, based on more recent evidence (Krewski and others, 2009; Lepeule and others, 2012; Industrial Economics Incorporated, 2006), US EPA now assumes a 10 microgram/cubic meter increase in $PM_{2.5}$ concentrations raises all pollution-related mortality risks by 10.6 percent (Chapter 5 of US EPA, 2011).

An important question is whether these findings—which are based on evidence for the United States, where $PM_{2.5}$ concentrations vary geographically by about 5–30 micrograms/cubic meter—apply to other regions. The assumptions used here are based on a best statistical fit for each of the four pollution-related illnesses of various model runs for different regions and different types of studies in Burnett and others (2013).[21] The resulting coefficients indicate that each 10 microgram/cubic meter increase in $PM_{2.5}$ concentrations increases the risk of all pollution-related mortality (averaged worldwide) by 9.8 percent. Although Burnett and others (2013) provide a state-of-the-art review of the limited number of studies, much more research is needed to improve the understanding of the complex relationship between pollution and mortality (especially for chronic illness); in the meantime, the health impacts calculated here should be viewed very cautiously.

A further caveat is that evidence suggests additional pollution exposure may, paradoxically, have significantly weaker impacts on mortality risk in regions where pollution concentrations are already very high, as the human body becomes progressively "saturated" with pollution (Burnett and others, 2013; Goodkind and others, 2012; Health Effects Institute, 2013). In other words, although the concentration response function appears to be approximately linear in pollution concentrations up to some point (an extra microgram/cubic meter of $PM_{2.5}$ has the same impact on mortality rates regardless of the initial pollution concentration level), eventually it may flatten out, that is, an extra microgram/cubic meter of $PM_{2.5}$ has a diminishing impact on elevating mortality rates, the higher the initial $PM_{2.5}$ concentration. However, as discussed in Box 3.3 of Chapter 3, the corrective fuel tax calculations abstract from this complication, on the assumption that implementing efficient taxes would have a large enough impact on emissions to lower pollution concentrations into the region where the concentration response function is approximately linear.[22]

Valuing Mortality Risks

Health risk valuation is highly controversial. Many people are uncomfortable with the idea of assigning values to the lives saved from policy interventions. Nonetheless, policymakers should still consider methodologies that have been developed for this exact purpose, despite the implication—unpalatable to some—that people with lower incomes are willing to sacrifice a smaller amount of their consumption to reduce health risks than people with much higher incomes.

In reality, people are constantly trading off money and mortality risk in a variety of decisions on a daily basis (e.g., when deciding whether to pay extra for a safer vehicle or to accept a higher-paying but riskier job like cleaning skyscraper windows). Economic studies attempt to measure these trade-offs, and a consistent

---

[21]Burnett and others (2013) bring together evidence from studies on mortality risks and exposure to ambient air pollution, emissions from solid cooking fuel, second-hand tobacco smoke, and active smoking. In the latter three cases, exposures were converted into estimated annual $PM_{2.5}$ exposure equivalents.

[22]This approach seems reasonable, based on Figure 1 in Burnett and others (2013), where saturation effects (at least for strokes) become especially pronounced only when $PM_{2.5}$ concentrations approach 100 micrograms/cubic meter. At least on a nationwide average basis, $PM_{2.5}$ concentrations are well below this level for typical countries (see Figure 2.9) and this is at current (rather than efficient) fuel tax levels.

finding across a broad range of countries is that mortality risk values generally rise with per capita income (OECD, 2012).

Methodological approaches for valuing mortality risks—or more precisely, the value per premature death avoided—are discussed below, along with empirical evidence, and the possible implications for different countries. Although not all governments will endorse this approach, the implications for corrective fuel taxes of alternative risk values are easily inferred from the results and accompanying spreadsheets by appropriate proportionate changes in the local pollution damage.

Methodological approaches

Two distinct approaches are often used to assess people's "willingness to pay" to reduce mortality risk. A third approach—generally less preferred by economists—based on valuing losses in human capital is discussed in Box 4.2.

The "revealed preference" method uses observed market behavior to assess mortality risk values, most usually by inferring a person's willingness to accept lower wages in return for a job with lower fatality risk (given other characteristics of jobs and workers). In contrast, the "stated preference" method relies on responses to questionnaires, most usually contingent-valuation studies in which people are asked direct questions about their money and risk trade-offs.

A potential drawback of revealed preference studies based on labor market data is that they focus on relatively healthy, average-age workers and on immediate accidental death in the workplace. Risks from pollution-related mortality—which

BOX 4.2

The Human Capital Approach

The human capital approach to valuing mortality risk does not (unlike willingness-to-pay approaches) measure people's own valuation of these risks—instead it focuses on measuring productivity losses from premature mortality. Traditionally this approach has been applied to lost years of working-age life, with a person's annual productivity proxied by market wages or per capita GDP, and productivity losses across future years discounted back to the present.

However, the human capital approach may undervalue the full economic cost of premature mortality in several respects. For example, the value of lost nonwork time (i.e., time in retirement and leisure time while working age) is often excluded. And people's valuation of pain and suffering before death are also excluded, as is grief to surviving family members. For these reasons, economists generally prefer willingness-to-pay approaches.[1]

---

[1] For comparison, in World Bank and State Environmental Protection Agency of China (2007), the costs of air and water pollution in China are about twice as high using the willingness-to-pay measures of mortality risk compared with the human capital measure.

primarily affects seniors and results from longer-term risk exposure—might be valued somewhat differently.

Stated preference studies can avoid these problems through choice of a sample that is more representative of the at-risk populations and through questions about specific hazards, such as cancer, posed by air pollution. The main concern with stated preference studies is that they are hypothetical—whether survey respondents would actually behave the way they say they would when confronted with risk-money trade-offs in the marketplace is unclear, leaving open the question of how accurately they describe people's actual trade-offs.

Both approaches focus on the costs to the individual (and grief to family members) from mortality risk and omit broader costs borne by third parties, such as medical costs. However, these broader costs may be small relative to the value of mortality risks to individuals; for example, when avoided medical costs later in the life cycle (from premature mortality) are subtracted from higher short-term treatment costs, the net medical burden may be relatively modest.

Empirical evidence

The starting value for mortality risk valuation used in this analysis, and its extrapolation to other countries, is based on a widely peer reviewed study by OECD (2012). This extrapolation accounts for differences in per capita income across countries but not, for reasons discussed in Box 4.3, for other factors, such as age.

BOX 4.3

Determinants Other than Income of Mortality Risk Valuation

OECD (2012) discusses several non-income-related factors that might cause mortality valuation to differ across countries, but in each case concludes that available evidence is not sufficiently conclusive to make adjustments.

With regard to population characteristics, conceivably the average age of the at-risk population matters, but whether, on balance, this has a positive or negative effect on mortality risk valuation is unclear. On the one hand, older individuals should have lower willingness to pay to reduce mortality risk given that they have fewer years of life left. Offsetting this, however, is that they might be wealthier and therefore have higher willingness to pay, compared with younger people, to increase expected longevity by a given amount. Some studies suggest there is little or no net effect of age on people's valuation of mortality risk, whereas others suggest a modest decrease at older ages (Krupnick, 2007; Chestnut, Rowe, and Breffle, 2004; Alberini and others, 2004; Hammitt, 2007). Two expert panels in the United States have recommended against age-related adjustments to mortality valuation (Cropper and Morgan, 2007; National Research Council, 2008), and the U.S. Environmental Protection Agency has, for now, abandoned analyses with these adjustments.

Even more unsettled is the appropriate value to apply to child mortality because children have not been the subject of revealed and stated preference studies. As noted, child mortality is excluded from the damage estimates in this book.

Evidence on whether healthier populations are willing to pay more to extend longevity than less healthy populations is similarly inconclusive (Krupnick and others, 2000). Unhealthy people may gain less enjoyment from living longer, but if they also gain less enjoyment from consumption, they may be willing to give up more consumption to prolong life. People in different countries may also have different preferences for trade-offs between consumption goods and mortality risks (perhaps because of cultural factors), but again there is no solid evidence on which to base an adjustment. Definitive evidence is also lacking on whether pollution-related risks (e.g., elevated cancer risk) are valued differently from accident risks, such as the risk of immediate death in a car accident.

Starting value for mortality risk reduction. In OECD (2012), the central case recommendation is to value mortality risks in OECD countries as a group at $3 million per life saved, in 2005 U.S. dollars.

This amount (which is updated below) was obtained from an extensive statistical analysis using several hundred stated preference studies applied to environmental, health, and traffic risks in a variety of countries (mostly Canada, China, France, the United Kingdom, and the United States). Stated preference studies were used because they have been conducted in numerous countries, while revealed preference studies have mainly been confined to the United States (which has ample labor market data). Stated preference studies tend to produce lower valuations than revealed preference studies; therefore, pollution damage estimates might be understated here.[23]

Income adjustment. The value for mortality risk per life for individual countries (denoted $V_{country}$) is extrapolated from that for the OECD as a whole (denoted V_{OECD}), using the following equation:

$$V_{country} = V_{OECD} \left(\frac{I_{country}}{I_{OECD}} \right)^{\varepsilon} \quad (4.1)$$

In equation (4.1), $I_{country}$ and I_{OECD} denote real income per capita in a particular country and for the OECD, respectively. Relative per capita income is appropriately measured using purchasing power parity rather than market exchange rates because purchasing power parity, which takes the local price level into account, more accurately reflects people's ability to pay out of their income for local products or risk reductions. The income per capita figures are obtained from IMF (2013) and World Bank (2013).

The exponent ε in (4.1) measures how mortality risk values vary with income; specifically, it is the percentage change in the mortality value per 1 percent change

---

[23] Exactly why most stated preference studies imply lower mortality risk valuations than revealed preference studies remains a puzzle.

in real per capita income. Based on OECD (2012), the illustrative calculations in this analysis assume ε is 0.8.[24]

The $3 million mortality value for the OECD is updated to 2010 for inflation (using the average consumer price index for the OECD) and real income (using equation (4.1) and the ratio of per capita income in the OECD in 2010 to that in 2005) to give V_{OECD} = $3.7 million. This amount is then extrapolated to other countries, using equation (4.1) and the countries' relative per capita incomes for 2010.

A tricky issue is how to value mortality risks for people across the border in other countries. To keep the exercise tractable, the same mortality risk value for these people is used as for people in the country in which emissions are released. An alternative, perhaps more appealing, approach would be to use a weighted average mortality risk valuation, whereby each affected country's risk valuation is weighted by its share of deaths in total deaths caused by the source country's emissions. If a source country has high per capita income relative to neighboring countries, this approach would imply somewhat lower pollution damage estimates than obtained in this analysis and vice versa for emissions from countries with relatively low income. However, the differences in emissions damage estimated by the two approaches may not be large; for example, if 40 percent of the affected population resides in other countries and mortality risks for these countries are 25 percent lower than the source country for the emissions, emissions damage will be 10 percent lower (a notable though not dramatic amount) compared with the approach taken here.

Implied mortality risk valuations

Figure 4.2 shows the implied mortality risk values for 20 selected countries. Mortality values per death are highest in the United States at $4.9 million and are less than $1 million in India, Indonesia, and Nigeria. To reemphasize, these values are purely illustrative—as shown below, if mortality values in all countries were set at the OECD average, the corrective tax estimates for relatively low-income countries would increase considerably.

The illustrated mortality values in this analysis differ quite a bit from values used at various points in different government studies. However, as shown by the examples in Table 4.1, there appears to be no systemic pattern to these differences. The values for the United States, Canada, and Germany used here are much lower than in government studies for these countries, but the converse applies in other cases. At any rate, the purpose is not to pass judgment on government practices but simply to obtain, for illustrative purposes, a consistently estimated set of cross-country mortality risk values.

Air Pollution Damage Estimates

Selected estimates of local air pollution damage per ton of emissions are discussed in this section. (Damage per unit of fuel is discussed in Chapter 6.) Annex

---

[24]Alan Krupnick, a leading expert on the issue, recommended a value of ε = 0.5, which would significantly increase pollution damage for relatively low-income countries.

Figure 4.2 Value of Mortality Risk, Selected Countries, 2010

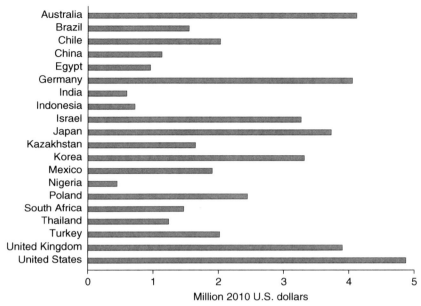

Source: Authors' calculations.
Note: Figure shows the value assigned to an individual's premature death caused by pollution.

TABLE 4.1
Examples of Mortality of Risk Valuations Used in Previous Government Studies

Country	Type of mortality risk	Year of study	Government value from study		Value used here relative to government value
			Local currency (thousands)	2010 US$ (thousands)	
Australia	General	2007	$A3,500	2,511	1.64
Austria	Transport	2009	€2,837	3,382	1.27
Canada	Transport	2006	Can$6,110	5,354	0.78
Denmark	Transport	2012	DKr16,070	1,769	2.43
France	Transport	2010	€1,360	1,503	2.50
Germany	Pollution	2009	€1,000–3,000	1,203–3,608	1.12–3.37
New Zealand	Transport	2009	$NZ3,500	2,179	1.54
Sweden	Transport	2010	SKr23739–31,331	2,558–3,377	1.24–1.64
United Kingdom	Transport	2000	£1,145	2,111	1.85
United States	Pollution	2006	US$7,400	8,007	0.61

Sources: Government websites and personal communications with government officials.

Table 4.2.1 provides the full set of estimates by emissions type, emissions source, and country.

Figure 4.3 shows estimated damage per metric ton for SO_2 from coal plants for selected countries. The range of the damage estimates is striking.

The United States, with damage of about $17,000/ton (in 2010 dollars), is an intermediate case. Damage estimates are much higher (about $35,000–$39,000/ton)

Figure 4.3 Damage from Coal Plant Sulfur Dioxide (SO$_2$) Emissions, Selected Countries, 2010

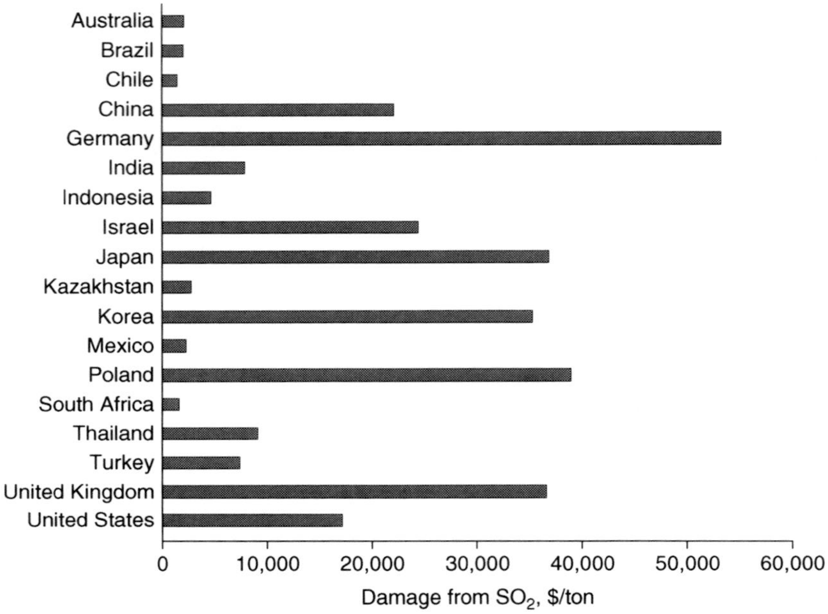

[Bar chart showing Damage from SO$_2$, $/ton, for countries: Australia, Brazil, Chile, China, Germany, India, Indonesia, Israel, Japan, Kazakhstan, Korea, Mexico, Poland, South Africa, Thailand, Turkey, United Kingdom, United States. X-axis: 0 to 60,000.]

Source: Authors' calculations.

in Japan, Poland, Korea, and the United Kingdom, and higher still (about $53,000/ton) in Germany, reflecting much higher population exposure to power plant emissions, which more than outweighs any influence of lower mortality values for these countries.

Conversely, as the result of a combination of lower population exposure and lower mortality risk values, Australia, Brazil, Chile, Kazakhstan, Mexico, and South Africa have dramatically lower damage values (about $1,500–$3,000/ton). For example, premature mortality per ton of emissions in Australia is just 15 percent of that for the United States.

Damage for China is about $22,000/ton. Although the illustrated mortality risk value for China is only 23 percent of that for the United States, this lower risk value is more than offset by an average population exposure to emissions that is six times as high.[25]

---

[25]The findings for China seem to be broadly in line with a far more sophisticated assessment of lives saved from an SO$_2$ control policy in Nielsen and Ho (2013). If the estimated acute and chronic deaths avoided in Nielsen and Ho (2013) are all attributed to the reductions in SO$_2$ emissions, about 25 lives are saved per kiloton reduction in SO$_2$ emissions, though this is an overstatement because some of the deaths avoided are due to indirect reductions in other pollutants (and Nielsen and Ho are careful to emphasize large uncertainties associated with these estimates). The calculations for this chapter suggest about 17 lives saved in China per kiloton reduction in SO$_2$ emissions.

Figure 4.4 illustrates damage per ton of SO_2 from coal combustion for all countries. Damage per ton is highest in European countries where both per capita income and population density are relatively high; for countries in North and South America, Asia, and Oceania damage per ton generally takes intermediate values. For Africa, many countries do not use coal, and for those that do, data limitations often preclude damage estimates.

On a per ton basis damage from primary coal plant emissions of $PM_{2.5}$ is about 25 percent larger than damage from SO_2. This is a broadly consistent finding across countries, so the relative pattern of damage across countries is similar to that for SO_2. Damage for NO_x from coal plants also follows the same broadly similar pattern across countries, though in absolute terms damage per ton from NO_x is about 20–50 percent lower than for SO_2 (mainly because NO_x is less prone to forming $PM_{2.5}$). Damage per ton from NO_x emissions from natural gas plants (essentially the only type of local emissions from these plants) is generally similar to the NO_x damage from coal plants.

Figure 4.5 shows the damage per ton from ground-level NO_x emissions (estimated for vehicles but also applied to home heating). Again, there are significant cross-country differences; for example, estimated damage exceeds $5,000/ton in Germany, Japan, Korea, and the United States but is less than $1,000/ton in India, Indonesia, Nigeria, South Africa, and Thailand. The relative differences are, however, smaller than for power plant emissions. Ground-level emissions tend to remain locally concentrated, therefore the large average distance between cities, such as in the United States, or the coastal location of cities, such as in Australia, reduce population exposure to a much lesser extent than for power plant emissions. Consequently, damage per ton in the United States is much closer to that in Germany, and in Australia is closer to typical European countries, as shown in Figure 4.5 compared with the relative damage for power plant emissions in Figure 4.3.

Robustness Checks

The assumptions about how pollution exposure affects health are based on state-of-the-art evidence from the Global Burden of Disease project—though this evidence is far from definitive—and Chapter 6 notes how corrective tax estimates vary with alternative values for mortality risk.

This subsection focuses on other issues relevant for tall smokestack emissions that are dispersed over great distances. The first issue is the reasonableness of the air pollution model (based on Zhou and others, 2006) implicitly underlying the intake fractions for China, and from which intake fractions for other countries are extrapolated. The second issue is how and to what extent the failure to capture cross-country differences in meteorology and related factors might bias damage estimates from the intake fraction approach.

These issues are examined by comparing selected results from the intake fraction approach with those from the TM5-Fast Scenario Screening Tool (FASST).[26]

---

[26] Simulations using this tool were conducted for this book by Nicholas Muller.

Figure 4.4 Damage from Coal Plant Sulfur Dioxide (SO$_2$) Emissions, All Countries, 2010

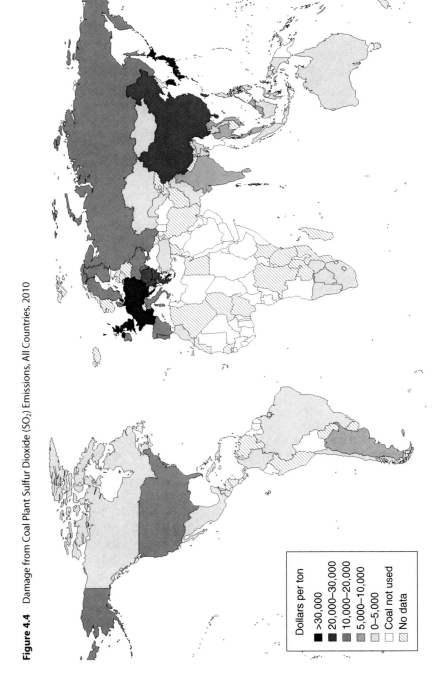

Source: Authors' calculations.

Figure 4.5 Damage from Ground-Level Nitrogen Oxide (NO$_x$) Emissions, Selected Countries, 2010

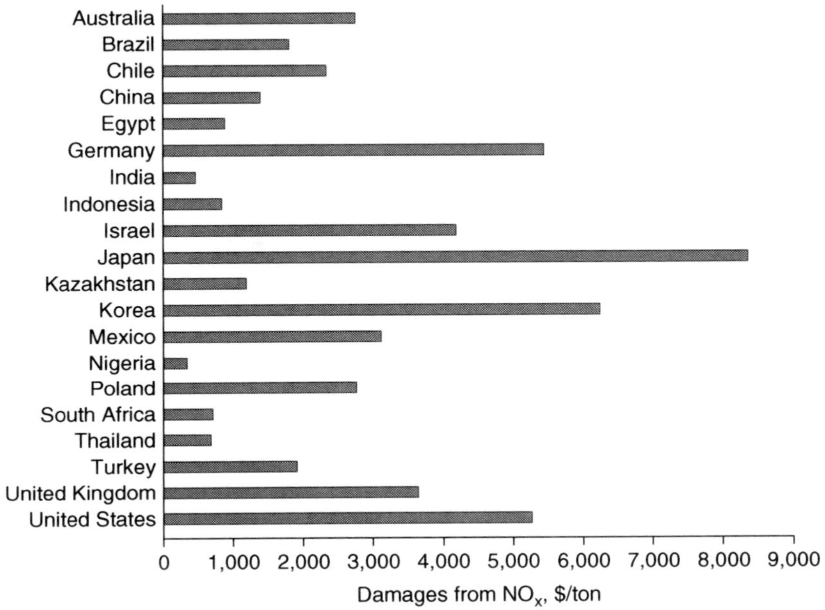

Source: Authors' calculations.

This tool (described in Annex 4.3) provides a simplified representation of how pollution concentrations in different regions change in response to additional emissions, and links these changes to population exposure and health impacts. The parameters underlying the air quality component of the model are chosen so that it yields predictions consistent with those from a highly sophisticated model of regional air pollution formation developed by the UN Environment Program (UNEP, 2011).

Unlike in the intake fraction approach, cross-country damage estimates from TM5-FASST capture regional differences in meteorology, ammonia concentration, and other factors. However, the estimation of population exposure is averaged over large areas—the world is divided into 51 regions—which understates population exposure if, as seems likely, power plants are located in areas with higher population density than the regional average.[27] Insofar as possible, other inputs to TM5-FASST—particularly baseline mortality rates by region and disease, impacts of additional PM$_{2.5}$ exposure on mortality rates, and the local valuation of mortality risks—are chosen to be consistent with the intake fraction approach, to facilitate a cleaner comparison of results.

---

[27]For example, one region is the whole of China, another the United States, and another combines Angola, Botswana, Malawi, Mozambique, Namibia, Zambia, and Zimbabwe.

With regard to the reasonableness of Zhou and others' (2006) air pollution model, TM5-FASST estimates SO_2 damage per ton for China to be about $12,000, or slightly more than half of the damage estimate from Zhou and others' (2006) intake fraction approach. Some of this difference reflects, as just noted, differences in population exposure, but some also likely reflects differences in assumptions about the impact of emissions on air quality. Unfortunately, it is not possible to make a definitive judgment about which air quality model is the more realistic.

With regard to the meteorology issue, Figures 4.6 and 4.7 show SO_2 damage per ton for selected countries expressed relative to damage per ton for China from the intake fraction approach and TM5-FASST, respectively. To the extent there are differences in relative damages for particular countries between the two figures, this suggests that differences in meteorological factors between that country and China play a potentially significant role. This does not appear to be a major concern for some countries—for example, the two approaches suggest damage per ton for Japan is 62–67 percent higher than for China, and for the United States damage is 22–24 percent lower than for China. But there are some exceptions; for example, relative damage for Israel, Poland, and the United Kingdom from the intake fraction approach is substantially higher than from

Figure 4.6 Estimated SO_2 Damage Relative to China Using the Intake Fraction Approach, 2010

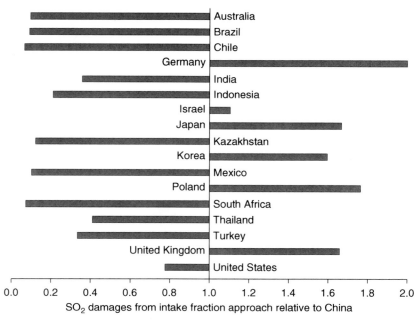

SO_2 damages from intake fraction approach relative to China

Source: Authors' calculations.

Figure 4.7 Estimated SO_2 Damage Relative to China Using the TM5-FASST Model, 2010

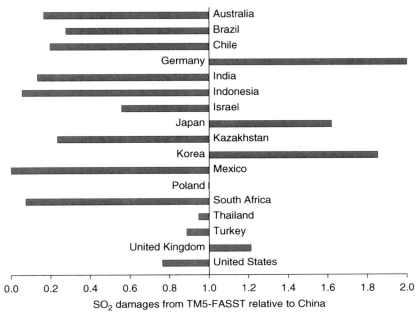

Source: Authors' calculations.

TM5-FASST,[28] and vice versa for Thailand and Turkey. In short, meteorological factors can significantly alter damage estimates in certain cases, though both the sign and scale of these effects are country specific.

Expressing Damage per Unit of Fuel

To assess efficient taxes on fuel use, the damage expressed per ton of emissions needs to be converted into damage per unit of fuel, or per unit of energy, using appropriate emissions factors. These factors relate the amount of emissions, such as SO_2, released into the atmosphere to combustion of a particular fuel, such as natural gas, in a particular activity, for example, power generation. The Greenhouse Gas and Air Pollution Interactions and Synergies (GAINS) model, developed by the International Institute for Applied Systems Analysis (IIASA), was used to estimate these factors.[29] See Box 4.4 for more details.

---

[28] A possible explanation is that winds blowing from west to east partially transport pollution away from these densely populated countries to less densely populated countries (like Scandinavia and Ukraine), which is taken into account in Figure 4.7 but not in Figure 4.6.

[29] For information about the GAINS model, see http://gains.iiasa.ac.at and IIASA (2013). The help of Fabian Wagner in calculating all the emission factors is gratefully acknowledged.

BOX 4.4

Emissions Factors from the GAINS Model

The Greenhouse Gas and Air Pollution Interactions and Synergies (GAINS) model estimates country-specific emissions factors for carbon and local air emissions for different fossil fuels used in different sectors of the economy. The estimates are reported in kilotons of pollutant per petajoule (heat content) of fuel input, though they could be expressed in emissions per unit of weight or volume (by multiplying heat content per unit of weight or volume using the GAINS data). Two calculations are performed.

First, uncontrolled emissions factors (denoted EF_U) are calculated from the basic properties of the fuel and combustion processes (Amann and others, 2011; Cofala and Syri, 1998a, 1998b; Klimont and others, 2002). For example, as defined for SO_2 emissions, the emissions factor is calculated by:

$$EF_U = \frac{sc}{hv} \times (1 - sr). \qquad (4.4.1)$$

In equation (4.4.1) sc is the sulfur content per unit of weight, hv is the heat value per unit of weight, and sr is the sulfur retention fraction (the portion of sulfur that is retained in ash rather than released into the atmosphere).

Second, various controlled emissions factors (denoted EF_C) are calculated, where applicable, for emissions sources employing an abatement technology denoted t (e.g., a particular type of scrubber, or hotter boiler that reduces NO_x emissions), from the following formula:

$$EF_C = EF_U \times (1 - re_t). \qquad (4.4.2)$$

In equation (4.4.2) re_t represents the fraction of emissions that are abated (that would otherwise be released into the atmosphere) as a result of technology t. Where specific regulations (technology mandates, emission rate standards) exist and are enforced, GAINS calculates the controlled emissions factors based on the regulation. GAINS can be used to calculate three emissions factors—an average over sources with controls (taking into account the potential application rates of alternative control technologies t), an average over sources with and without controls, and an uncontrolled emissions rate.

The GAINS data for these calculations are detailed for some countries; for others, judgment is used to transfer estimates to countries for which data are not directly available.

Coal emissions factors (for CO_2, SO_2, NO_x, and primary $PM_{2.5}$) are defined relative to energy or heat content in petajoules (PJ) rather than tons of coal, given significant variation in energy content across different types of coal. Where relevant, the factors represent a weighted average across different coal types—in these cases, a more refined pricing system than the one estimated here would vary charges according to the emissions intensity of the particular coal type.

One emissions factor is obtained for carbon, for which opportunities for abating emissions at the point of combustion are presently very limited. For the local air pollutants from coal plants, three different emissions factors are distinguished for each pollutant. First is an uncontrolled emission rate. Second is an emission rate for a representative plant that has some control technology (e.g., an SO_2 scrubber). Third is the average emission rate across all existing plants with and without emissions control technologies. These factors are used to estimate three different taxes, for coal plants without and with emissions controls, and for the average coal plant at present, though crediting would provide strong incentives for all plants to use control technologies. In each case, the corrective tax is the product of the emissions factor for a pollutant and the damage per ton for that pollutant, which is then aggregated over all pollutants.

Similar procedures are used to obtain emissions factors and corrective taxes for natural gas. For power plants for which the local air pollution damages are small relative to those from coal, the focus is just on the emissions factor averaged across all plants with and without control technologies, whereas for household use of gas only uncontrolled emissions are relevant. Damage and corrective taxes are again expressed per unit of energy because emissions per unit of volume can vary significantly depending on gas pressure.

For mobile sources, emissions factors for CO_2, SO_2, NO_x, and $PM_{2.5}$ per liter are obtained for gasoline vehicles and diesel vehicles (the latter representing an average of light- and heavy-duty vehicles using diesel), in each case averaging across vehicles on the road with and without control technologies (fuel taxes by themselves do not encourage the adoption of control technologies).

There are several noteworthy points about the emissions factors:

First, carbon emissions factors for a particular fuel vary little across countries. However, the fuel products themselves vary significantly: per PJ of energy, natural gas, gasoline, and motor diesel generate about 59 percent, 73 percent, and 78 percent, respectively, of the carbon emissions generated by one PJ of coal.

Second, uncontrolled, average, and controlled SO_2 emissions factors for coal can vary greatly both within and across countries (see Figure 4.8). For example, on average, the SO_2 emissions per PJ for Japanese coal plants with no control technologies is only 30 percent of that for comparable U.S. plants, while in Israel the emission rate is about 70 percent greater than for the United States. Control technologies can dramatically reduce emissions, however; for example, SO_2 emission rates at U.S. coal plants with such technologies are 95 percent lower than the rates at plants without these technologies. Primary $PM_{2.5}$ emission rates for uncontrolled plants follow a pattern similar to those for SO_2, but the control technologies have an even more dramatic impact on reducing pollution.

Third, NO_x emission rates from plants without controls differ from rates for ground-level sources (depending, for example, on combustion temperature, which can affect the amount of nitrogen and oxygen sucked in from the ambient air), but the differences are not large.

Figure 4.8 SO$_2$ Emissions Rates at Coal Plants, 2010

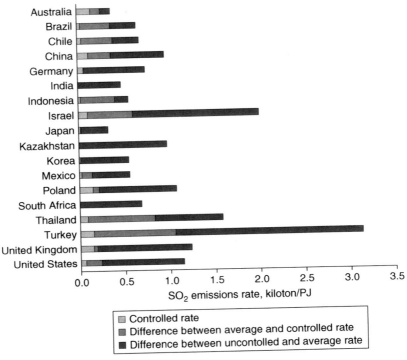

Source: Authors' calculations.
Note: PJ = petajoule. Controlled rate is the average emissions factor for plans with emissions controls. Uncontrolled rate is for plants with no controls. Average rate averages across plants with and without controls. For some countries (e.g., India, South Africa), no sulfur control technologies had been adopted as of 2010. In Germany the average and controlled rates are the same because all plants had some form of control.

SUMMARY

An illustrative value for CO_2 damage is taken from a recent study, although this value is subject to much debate.

To assess air pollution damage from power plant combustion of coal and natural gas, average population exposure to these emissions (which can be transported long distances) is estimated and combined with data on local mortality rates for pollution-related illness and evidence about how changes in pollution exposure affect these mortality rates. The most controversial step is monetizing these health effects, which, for illustration are inferred from OECD (2012), though the implications of alternative assumptions are transparent from the results. Estimation of pollution from motor vehicle fuels and other ground-level sources are constructed from studies about how much of these emissions are inhaled by people in different urban centers. These pollution damage estimates are then combined with data on local emissions factors for different fuels to derive environmental damage from fuel use (though emission rates vary considerably with the extent of use of control technologies).

ANNEX 4.1. REGIONAL CLASSIFICATIONS FOR MORTALITY RATES

Baseline mortality rates are obtained by combining regional average mortality rates for four pollution-related illnesses by age group with country-level data on the age structure of the population. The regional mortality data are from Burnett and others (2013), with the countries grouped into regional classifications as shown in Annex Table 4.1.1.

ANNEX TABLE 4.1.1

Country Classifications for Baseline, Pollution-Related Mortality Rates

Asia Pacific, High Income	Australasia	Europe, Western	Latin America, Andean	Sub-Sah. Africa, Central	Sub-Sah. Africa, West
Brunei	Australia	Andorra	Bolivia	Angola	Benin
Japan	New Zealand	Austria	Ecuador	Central African Republic	Burkina Faso
Korea, Rep. of	**Oceania**	Belgium	Peru	Dem. Republic of the Congo	Côte dIvoire
Singapore	Fiji	Switzerland		Rep. of Congo	Cabo Verde
Asia, Central	Fed. States of Micronesia	Cyprus	**Latin America, Central**	Gabon	Cameroon
Armenia	Kiribati	Germany	Colombia	Equatorial Guinea	Chad
Azerbaijan	Marshall Islands	Denmark	Costa Rica		Ghana
Georgia	Papua New Guinea	Spain	Guatemala		Guinea
Kazakhstan	Solomon Islands	Finland	Honduras	**Sub-Sah. Africa, East**	The Gambia
Kyrgyzstan	Tonga	France	Mexico	Burundi	Guinea-Bissau
Mongolia	Vanuatu	United Kingdom	Nicaragua	Comoros	Liberia
Tajikistan	Samoa	Greece	Panama	Djibouti	Mali
Turkmenistan		Ireland	El Salvador	Eritrea	Mauritania
Uzbekistan	**Europe, Central**	Iceland	Venezuela	Ethiopia	Niger
Asia, East	Albania	Israel		Kenya	Nigeria
China	Bulgaria	Italy	**Latin America, Southern**	Madagascar	Senegal
Korea, Dem. Rep. of	Bosnia and Herzegovina	Luxembourg	Argentina	Mozambique	Sierra Leone
Taiwan Province of China	Czech Republic	Malta	Chile	Mauritius	São Tomé and Príncipe
	Croatia	Netherlands	Uruguay	Malawi	Togo
Asia, South	Hungary	Norway		Rwanda	
Afghanistan	FYR Macedonia	Portugal	**Latin America, Tropical**	Sudan	
Bangladesh	Montenegro	Sweden	Brazil	Somalia	
Bhutan	Poland		Paraguay	Seychelles	
India	Romania	**North America, High Income**		Tanzania	
Nepal	Serbia	Canada	**Middle East and North Africa**	Uganda	
Pakistan	Slovak Republic	United States	United Arab Emirates	Zambia	
	Slovenia	**Caribbean**	Bahrain		
Asia, Southeast		Antigua and Barbuda	Algeria	**Sub-Sah. Africa, Southern**	
Indonesia	**Europe, Eastern**	The Bahamas	Egypt	Botswana	
Cambodia	Belarus	Belize	Iran	Lesotho	
Lao P.D.R.	Estonia	Barbados	Iraq	Namibia	
Sri Lanka	Lithuania	Cuba	Jordan	Swaziland	
Maldives	Latvia	Dominica	Kuwait	South Africa	
Myanmar	Moldova	Dominican Republic	Lebanon	Zimbabwe	
Malaysia	Russia	Grenada	Libya		
Philippines	Ukraine	Guyana	Morocco		
Thailand		Haiti	Oman		
Timor-Leste		Jamaica	Palestine		
Vietnam		Saint Lucia	Qatar		
		Suriname	Saudi Arabia		
		Trinidad and Tobago	Syria		
		St. Vincent and the Grenadines	Tunisia		
			Turkey		
			Yemen		

Source: Burnett and others (2013).

ANNEX 4.2. AIR POLLUTION DAMAGE BY EMISSIONS AND COUNTRY

Annex Table 4.2.1 summarizes, by country, estimated air pollution damage by the type of emissions, and the source of those emissions.

ANNEX TABLE 4.2.1

Damage from Local Air Pollution, All Countries, $/ton of Emissions, 2010

	Sulfur dioxide			Nitrogen oxides			Primary fine particulate matter		
	$ per ton			$ per ton			$ per ton		
Country	coal	natural gas	ground level	coal	natural gas	ground level	coal	natural gas	ground level
North America									
Canada	3,908	8,994	13,284	2,757	5,917	2,712	4,887	11,480	349,161
Mexico	2,240	3,599	7,704	1,797	2,179	1,575	2,700	4,416	203,680
United States	17,132	18,978	17,005	12,472	12,092	3,468	21,402	23,294	445,484
Central and South America									
Argentina	8,328	4,928	7,553	3,475	2,375	1,532	10,420	6,167	193,736
Barbados	#na	26,387	**#na**	#na	12,166	**#na**	#na	33,014	**#na**
Bolivia	#na	343	355	#na	237	73	#na	410	9,624
Brazil	2,004	4,293	5,013	1,492	2,401	1,021	2,626	5,258	130,726
Chile	1,409	1,989	7,057	1,029	1,185	1,434	1,730	2,482	182,276
Colombia	1,867	1,648	6,180	1,162	1,084	1,265	2,307	2,047	164,342
Costa Rica	**#na**	**#na**	2,316	**#na**	**#na**	477	**#na**	**#na**	63,036
Cuba	**#na**	4,447	3,627	**#na**	3,033	743	**#na**	5,293	96,420
Dominican Republic	3,007	**#na**	3,475	1,694	**#na**	714	3,713	**#na**	93,597
Ecuador	#na	748	721	#na	454	148	#na	915	19,505
El Salvador	#na	**#na**	696	#na	**#na**	143	#na	**#na**	18,949
Guatemala	882	**#na**	417	501	**#na**	87	1,076	**#na**	11,721
Honduras	**#na**	**#na**	936	**#na**	**#na**	194	**#na**	**#na**	26,190
Jamaica	**#na**	**#na**	1,617	**#na**	**#na**	336	**#na**	**#na**	45,210
Nicaragua	**#na**	560	265	**#na**	374	55	**#na**	665	7,338
Panama	1,560	**#na**	1,581	1,079	**#na**	324	2,031	**#na**	42,080
Paraguay	#na	**#na**	825	#na	**#na**	170	#na	**#na**	22,594
Peru	359	1,415	2,435	290	593	498	447	1,767	64,499
St. Vincent and the Grenadines	#na	**#na**	**#na**	#na	**#na**	**#na**	#na	**#na**	**#na**
Suriname	#na	**#na**	649	#na	**#na**	133	#na	**#na**	17,461
Trinidad and Tobago	#na	2,883	**#na**	#na	1,977	**#na**	#na	3,553	**#na**
Uruguay	**#na**	3,151	2,184	**#na**	2,164	443	**#na**	3,773	55,994
Venezuela	**#na**	2,027	4,000	**#na**	1,203	811	**#na**	2,575	102,381
Europe									
Albania	**#na**	#na	4,927	**#na**	#na	1,023	**#na**	#na	137,666
Austria	41,004	41,889	12,951	31,812	31,666	2,664	51,736	53,150	350,052
Belgium	53,017	51,863	10,883	34,613	34,243	2,201	64,698	63,189	276,234
Bosnia and Herzegovina	**#na**	**#na**	5,556	**#na**	**#na**	1,157	**#na**	**#na**	156,869
Bulgaria	23,980	**#na**	7,536	19,472	**#na**	1,545	28,991	**#na**	201,479
Croatia	35,046	35,676	10,533	28,197	27,410	2,179	44,610	45,720	290,953
Cyprus	**#na**	#na	2,232	**#na**	#na	458	**#na**	#na	59,950
Czech Republic	56,034	55,308	9,670	40,836	41,184	1,982	69,818	68,676	258,025
Denmark	26,136	26,025	6,276	20,048	19,993	1,277	34,589	34,627	162,816
Finland	14,814	16,035	10,786	12,152	12,711	2,198	17,739	19,320	281,719

ANNEX TABLE 4.2.1
Damage from Local Air Pollution, All Countries, $/ton of Emissions, 2010

Country	Sulfur dioxide $ per ton			Nitrogen oxides $ per ton			Primary fine particulate matter $ per ton		
	coal	natural gas	ground level	coal	natural gas	ground level	coal	natural gas	ground level
France	33,555	37,779	15,908	24,511	27,670	3,239	41,725	46,003	414,075
Germany	53,192	56,125	20,082	35,624	36,603	4,115	65,936	69,514	535,454
Greece	20,699	20,734	8,028	16,843	16,213	1,657	25,562	25,570	219,970
Hungary	41,057	40,925	11,070	30,712	30,608	2,275	51,744	51,840	298,250
Iceland	#na	#na	3,855	#na	#na	781	#na	#na	98,626
Ireland	12,897	18,828	4,991	10,468	14,585	1,030	16,217	22,833	136,535
Italy	26,627	31,596	13,346	20,905	22,958	2,744	33,654	40,278	360,129
Luxembourg	#na	86,775	#na	#na	65,283	#na	#na	105,443	#na
Macedonia, FYR	16,736	17,560	5,832	13,541	14,096	1,206	20,686	21,656	160,515
Malta	#na	#na	#na	#na	#na	#na	#na	#na	#na
Montenegro	21,031	#na	4,205	17,103	#na	867	26,405	#na	114,743
Netherlands	53,065	50,535	13,357	35,421	34,581	2,723	65,304	62,168	349,477
Norway	#na	17,667	35,210	#na	14,920	7,194	#na	23,495	928,330
Poland	38,887	35,828	9,468	28,429	27,749	1,955	49,082	45,043	259,582
Portugal	12,221	12,533	6,383	9,265	9,355	1,318	14,755	15,177	175,156
Romania	26,813	27,895	7,995	21,377	21,041	1,659	33,293	34,439	223,169
Serbia	24,142	24,194	6,728	18,319	18,274	1,393	30,381	30,841	186,463
Slovak Republic	42,444	46,050	7,275	32,616	33,770	1,508	53,469	58,463	202,158
Slovenia	52,466	52,388	10,936	39,744	39,419	2,273	67,044	66,807	307,217
Spain	16,871	19,270	19,055	13,364	14,498	3,897	20,852	23,980	504,326
Sweden	17,058	19,702	16,370	13,005	15,757	3,333	21,281	25,956	426,238
Switzerland	#na	46,015	11,919	#na	34,809	2,443	#na	57,827	317,909
Turkey	7,341	9,611	5,264	5,746	6,507	1,081	9,146	11,858	141,362
United Kingdom	36,577	40,069	12,325	22,857	27,378	2,518	45,415	48,658	324,687
Eurasia									
Armenia	#na	7,411	3,020	#na	5,584	622	#na	9,156	82,228
Azerbaijan	#na	8,462	3,498	#na	6,417	726	#na	10,520	97,516
Belarus	#na	26,576	15,038	#na	21,381	3,080	#na	33,671	400,285
Estonia	#na	28,605	8,435	#na	22,914	1,733	#na	34,958	226,999
Georgia	#na	6,049	2,762	#na	4,613	573	#na	7,525	77,102
Kazakhstan	2,668	6,107	3,104	2,225	5,306	644	3,184	7,588	86,461
Kyrgyzstan	1,934	#na	654	1,518	#na	137	2,328	#na	19,010
Latvia	23,252	28,935	10,572	19,784	23,459	2,174	29,743	36,413	285,607
Lithuania	#na	34,985	13,522	#na	27,769	2,782	#na	44,700	365,862
Russia	17,562	22,105	32,383	12,508	14,317	6,637	21,525	27,714	863,732
Tajikistan	#na	#na	418	#na	#na	88	#na	#na	12,393
Turkmenistan	#na	5,775	1,632	#na	4,770	340	#na	6,978	46,015
Ukraine	17,851	16,728	6,377	13,593	12,690	1,311	22,086	20,497	171,913
Uzbekistan	3,451	2,797	659	2,552	2,162	138	4,175	3,359	19,116
Middle East									
Bahrain	#na	7,161	2,451	#na	5,303	498	#na	8,563	63,360
Iran	#na	5,066	3,956	#na	3,694	813	#na	6,171	106,587
Iraq	#na	1,171	857	#na	877	176	#na	1,482	23,197
Israel	24,369	24,926	11,652	15,717	15,759	2,364	29,482	30,226	299,185
Jordan	#na	2,429	1,113	#na	1,643	227	#na	2,975	29,144
Kuwait	#na	#na	9,771	#na	#na	1,976	#na	#na	247,625
Lebanon	#na	7,253	2,080	#na	4,753	423	#na	9,202	53,922
Oman	#na	7,088	3,095	#na	6,022	634	#na	8,028	82,631
Qatar	#na	16,731	7,246	#na	13,738	1,465	#na	19,600	183,468
Saudi Arabia	#na	4,895	4,651	#na	3,641	949	#na	6,018	121,849
Syria	#na	2,829	1,404	#na	1,864	291	#na	3,612	38,929
United Arab Emirates	#na	6,431	3,019	#na	4,845	615	#na	7,578	78,782

(Continued)

ANNEX TABLE 4.2.1 (Continued)
Damage from Local Air Pollution, All Countries, $/ton of Emissions, 2010

	Sulfur dioxide			Nitrogen oxides			Primary fine particulate matter		
	$ per ton			$ per ton			$ per ton		
Country	coal	natural gas	ground level	coal	natural gas	ground level	coal	natural gas	ground level
Africa									
Algeria	#na	3,442	1,834	#na	2,381	376	#na	4,242	49,099
Angola	#na	465	1,320	#na	312	273	#na	567	36,392
Benin	#na	#na	75	#na	#na	16	#na	#na	2,122
Botswana	1,007	656	680	798	556	140	1,238	879	18,629
Burkina Faso	#na	#na	68	#na	#na	14	#na	#na	2,027
Burundi	#na	#na	16	#na	#na	3	#na	#na	494
Cameroon	#na	312	419	#na	254	87	#na	391	11,732
Cabo Verde	#na	#na	#na	#na	#na	#na	#na	#na	#na
Central African Republic	#na	#na	130	#na	#na	27	#na	#na	3,745
Comoros	#na	#na	#na	#na	#na	#na	#na	#na	#na
Congo, Rep. of	#na	66	87	#na	52	18	#na	77	2,378
Côte d'Ivoire	#na	312	289	#na	197	60	#na	391	8,096
Egypt	#na	5,288	1,912	#na	2,764	399	#na	6,460	54,506
Ethiopia	#na	#na	70	#na	#na	15	#na	#na	2,114
Gambia, The	#na	#na	73	#na	#na	15	#na	#na	2,017
Ghana	#na	270	117	#na	197	24	#na	344	3,273
Guinea-Bissau	#na	#na	81	#na	#na	17	#na	#na	2,315
Kenya	#na	234	90	#na	173	19	#na	289	2,683
Liberia	#na	#na	171	#na	#na	35	#na	#na	4,814
Libya	#na	2,470	1,296	#na	1,942	265	#na	2,952	34,272
Madagascar	#na	#na	81	#na	#na	17	#na	#na	2,371
Malawi	#na	148	38	#na	91	8	#na	#na	1,164
Mali	#na	#na	56	#na	#na	12	#na	#na	1,621
Mauritius	438	#na	#na	206	#na	#na	545	#na	#na
Morocco	1,540	1,762	1,563	930	1,085	324	1,901	2,167	43,251
Mozambique	#na	#na	44	#na	#na	9	#na	#na	1,303
Namibia	202	#na	281	167	#na	59	233	#na	8,111
Niger	#na	#na	28	#na	#na	6	#na	#na	844
Nigeria	#na	714	535	#na	425	111	#na	887	15,051
Rwanda	#na	#na	51	#na	#na	11	#na	#na	1,545
São Tomé and Príncipe	#na	#na	#na	#na	#na	#na	#na	#na	#na
Senegal	134	#na	112	71	#na	23	164	#na	3,188
Seychelles	#na	#na	#na	#na	#na	#na	#na	#na	#na
Sierra Leone	#na	#na	68	#na	#na	14	#na	#na	1,959
South Africa	1,602	2,550	1,690	1,031	1,219	349	1,905	3,154	46,284
Sudan and South Sudan	#na	207	100	#na	171	21	#na	239	2,934
Swaziland	#na	#na	#na	#na	#na	#na	#na	#na	#na
Tanzania	#na	175	116	#na	115	24	#na	221	3,429
Togo	#na	272	44	#na	187	9	#na	345	1,261
Tunisia	#na	3,925	1,834	#na	2,952	378	#na	4,758	49,730
Uganda	#na	#na	44	#na	#na	9	#na	#na	1,340
Zambia	#na	#na	84	#na	#na	18	#na	#na	2,430
Zimbabwe	51	#na	50	41	#na	10	65	#na	1,435
Asia and Oceania									
Afghanistan	#na	866	186	#na	642	39	#na	1,077	5,545
Australia	2,098	2,136	9,220	1,129	900	1,873	2,632	2,698	238,099
Bangladesh	6,057	6,131	1,757	4,082	3,757	371	7,181	7,430	51,932
Bhutan	#na	#na	#na	#na	#na	#na	#na	#na	#na
Brunei	#na	10,797	#na	#na	9,274	#na	#na	12,225	#na
Cambodia	#na	#na	486	#na	#na	103	#na	#na	14,655
China	22,045	25,577	4,422	15,530	16,605	920	27,609	32,238	124,441
Fiji	#na	#na	#na	#na	#na	#na	#na	#na	#na

ANNEX TABLE 4.2.1

Damage from Local Air Pollution, All Countries, $/ton of Emissions, 2010

| | Sulfur dioxide | | | Nitrogen oxides | | | Primary fine particulate matter | | |
| | $ per ton | | | $ per ton | | | $ per ton | | |
Country	coal	natural gas	ground level	coal	natural gas	ground level	coal	natural gas	ground level
Hong Kong SAR	82,580	72,288	#na	53,207	49,085	#na	103,759	91,246	#na
India	7,833	6,837	1,093	5,683	4,762	230	9,773	8,549	32,075
Indonesia	4,617	5,627	2,159	2,492	2,699	449	5,636	6,936	60,669
Japan	36,786	47,176	31,548	24,230	24,772	6,405	44,381	57,309	812,178
Kiribati	#na	#na	**#na**	#na	#na	**#na**	#na	#na	**#na**
Korea, Rep. of	35,228	34,688	20,862	25,439	25,375	4,253	46,054	45,507	545,623
Malaysia	6,525	6,104	4,028	4,360	4,273	826	7,891	7,406	107,824
Maldives	#na	#na	**#na**	#na	#na	**#na**	#na	#na	**#na**
Mongolia	3,138	#na	2,253	2,736	#na	463	3,498	#na	60,870
New Zealand	1,568	1,296	2,508	479	396	510	1,981	1,637	65,153
Pakistan	2,254	2,902	630	1,698	2,075	132	2,942	3,663	18,290
Papua New Guinea	#na	**#na**	91	#na	**#na**	19	#na	**#na**	2,777
Philippines	3,372	4,426	1,393	1,969	2,246	290	4,053	5,377	39,237
Samoa	#na	#na	**#na**	#na	#na	**#na**	#na	#na	**#na**
Singapore	**#na**	21,698	42,652	**#na**	13,439	8,617	**#na**	27,223	1,077,044
Sri Lanka	4,262	#na	410	3,258	#na	87	5,068	#na	12,500
Taiwan Province of China	46,892	49,692	**#na**	35,615	36,445	**#na**	59,253	63,012	**#na**
Thailand	9,036	9,067	2,013	6,941	6,087	423	10,886	11,105	58,683
Vietnam	5,823	3,274	1,416	4,060	2,028	298	7,243	3,989	41,622

Source: See main text.
Note: The table shows estimates of the local pollution health damage per ton from each of three pollutants, according to whether emissions are released from coal combustion, natural gas combustion at power plants, or natural gas and motor fuel consumption at ground level. Black #na = fuel not used; bold #na = data not available.

ANNEX 4.3. DETAILS ON THE TM5-FASST TOOL

The TM5-FASST tool estimates air pollution damage per ton for different types of emissions in several steps.

First, the baseline mortality rates for four pollution-related illnesses are calculated by region, according to equation (4.2):

$$RR(PM_{2.5}) = 1 + \alpha \times \Delta PM_{2.5}. \qquad (4.2)$$

RR denotes the risk of premature death from a particular illness relative to that in the baseline case (with current pollution concentration levels). *RR*–1 is therefore the proportionate change in the relative risk. $\Delta PM_{2.5}$ denotes the change in $PM_{2.5}$ concentrations relative to the initial situation. α is a parameter that is calibrated separately for each of the four pollution-related diseases to be consistent with the evidence in Burnett and others (2013).

The change in premature deaths for a change in $PM_{2.5}$ concentrations is given by equation (4.3):

$$(RR - 1) \times \textit{mortality rate} \times \textit{population}, \qquad (4.3)$$

in which *mortality rate* refers to the baseline rate and *population* is the exposed population (all those ages 25 and older). Both population data and mortality rate data are from the Institute for Health Metrics and Evaluation (IHME). Deaths are monetized using the same mortality values as discussed in the main text of this chapter.

Next, one ton of SO_2 emissions from a particular source is added and processed through an air quality model that links all emissions sources to $PM_{2.5}$ concentrations in the model's 51 different regions. The air quality model in the FASST tool is a simplified version of a far more sophisticated air quality model in UN Environment Programme (2011). The change in $PM_{2.5}$ concentrations in each region is then used to calculate changes in premature deaths based on equations (4.2) and (4.3), and the result is then monetized.

Averaged across the 20 countries considered in this chapter, the damage from SO_2 is $17,640/ton. Individual country estimates, relative to those from the intake fraction approach used here, are discussed in the main text of this chapter.

As a check on the above results, simulations were also run with an alternative specification for relative risk given by equation (4.4):

$$RR(PM_{2.5}) = 1 + \alpha(1 - e^{-\gamma \Delta PM_{2.5}^{\delta}}), \qquad (4.4)$$

in which parameters α, γ, and δ are calibrated for each of the four diseases to be consistent with Burnett and others (2013). However, the results are only moderately affected. For example, the average damage from SO_2 across the 20 countries is 15 percent smaller than when the linear functional form is used.

REFERENCES

Alberini, Anna, Maureen Cropper, Alan Krupnick, and Nathalie B. Simon, 2004, "Does the Value of a Statistical Life Vary with Age and Health Status? Evidence from the US and Canada," *Journal of Environmental Economics and Management,* Vol. 48, pp. 769–92.

Amann, Markus, Imrich Bertok, Jens Borken-Kleefeld, Janusz Cofala, Chris Heyes, Lena Höglund-Isaksson, Zbigniew Klimont, and others, 2011, "Cost-Effective Control of Air Quality and Greenhouse Gases in Europe: Modeling and Policy Applications," *Environmental Modelling and Software,* Vol. 26, pp. 1489–501.

Apte, Joshua S., Emilie Bornbrun, Julian D. Marshall, and William W. Nazaroff, 2012, "Global Intraurban Intake Fractions for Primary Air Pollutants from Vehicles and Other Distributed Sources," *Environmental Science & Technology,* Vol. 46, pp. 3415–23.

Bennett, D.H., T.E. McKone, J.S. Evans, W.W. Nazaroff, M.D. Margni, O. Jolliet, and K.R. Smith, 2002, "Defining Intake Fraction," *Environmental Science & Technology,* Vol. 36, No. 9, pp. 207–16.

Bosetti, Valentina, Sergey Paltsev, John Reilly, and Carlo Carraro, 2012, "Emissions Pricing to Stabilize Global Climate," in *Fiscal Policy to Mitigate Climate Change: A Guide for Policymakers,* edited by I.W.H. Parry, R. de Mooij, and M. Keen (Washington: International Monetary Fund).

Burnett, Richard T., C. Arden Pope, Majid Ezzati, Casey Olives, Stephen S. Lim, Sumi Mehta, Hwashin H. Shin, and others, 2013, "An Integrated Risk Function for Estimating the Global Burden of Disease Attributable to Ambient Fine Particulate Matter Exposure" (Unpublished; Ottawa, Ontario, Canada: Health Canada).

Chestnut, L.G., R.D. Rowe, and W.S. Breffle, 2004, "Economic Valuation of Mortality Risk Reduction: Stated Preference Approach in Canada," Report prepared for Health Canada (Boulder, Colorado: Stratus Consulting Inc).

Cofala, J., and S. Syri, 1998a, "Nitrogen Oxides Emissions, Abatement Technologies and Related Costs for Europe in the RAINS Model Database" (Laxenburg, Austria: International Institute for Applied Systems Analysis).

———, 1998b, "Sulfur Emissions, Abatement Technologies and Related Costs for Europe in the RAINS Model Database" (Laxenburg, Austria: International Institute for Applied Systems Analysis).

Cropper, Maureen, Shama Gamkhar, Kabir Malik, Alex Limonov, and Ian Partridge, 2012, "The Health Effects of Coal Electricity Generation in India," Discussion Paper No. 12–15 (Washington: Resources for the Future).

Cropper, Maureen, and M. Granger Morgan, 2007, "SAB Advisory on EPA's 'Issues in Valuing Mortality Risk Reduction,'" Memorandum from the Chair, Science Advisory Board, and the Chair, Environmental Economics Advisory Committee, to EPA Administrator Stephen L. Johnson. EPA-SAB-08–001.

European Commission (EC), 2008, *ExternE-Externalities of Energy: A Research Project of the European Commission* (Brussels: European Commission).

Gillingham, Robert, and Michael Keen, 2012, "Mitigation and Fuel Pricing in Developing Countries," in *Fiscal Policy to Mitigate Climate Change: A Guide for Policymakers,* edited by I.W.H. Parry, R. de Mooij and M. Keen (Washington: International Monetary Fund).

Goodkind, Andrew L., Jay S. Coggins, Timothy A. Delbridge, Milda Irhamni, Justin Andrew Johnson, Suhyun Jung, Julian Marshall, Bijie Ren, and Martha H. Rogers, 2012, "Prices vs. Quantities With Increasing Marginal Benefits," Discussion paper, Department of Applied Economics, University of Minnesota.

Griffiths, Charles, Elizabeth Kopits, Alex Marten, Chris Moore, Steve Newbold, and Ann Wolverton, 2012, "The Social Cost of Carbon: Valuing Carbon Reductions in Policy Analysis," in *Fiscal Policy to Mitigate Climate Change: A Guide for Policymakers,* edited by Ian W.H. Parry, Ruud de Mooij, and Michael Keen (Washington: International Monetary Fund).

Hammitt, James K., 2007, "Valuing Changes in Mortality Risk: Lives Saved Versus Life Years Saved," *Review of Environmental Economics and Policy,* Vol. 1, pp. 228–40.

Health Effects Institute, 2013, "Understanding the Health Effects of Ambient Ultrafine Particles," HEI Review Panel on Ultrafine Particles (Boston: Health Effects Institute).

Humbert, Sebastien, Julian D. Marshall, Shanna Shaked, Joseph V. Spadaro, Ryrika Nichioka, Philipp Preiss, Thomas E. McKone, Arpad Horvath, and Olivier Jolliet, 2011, "Intake Fraction for Particulate Matter: Recommendations for Life Cycle Impact Assessment," *Environmental Science & Technology*, Vol. 45, pp. 4808–16.

Industrial Economics, Incorporated, 2006, "Expanded Expert Judgment Assessment of the Concentration-Response Relationship between PM2.5 Exposure and Mortality," Final Report (Cambridge, Massachusetts: Industrial Economics, Incorporated). www.epa.gov/ttn/ecas/regdata/Uncertainty/pm_ee_report.pdf.

International Institute for Applied Systems Analysis (IIASA), 2013, "Greenhouse Gas-Air Pollution Interaction and Synergies Model" (Laxenburg, Austria: International Institute for Applied Systems Analysis). www.Gains.iiasa.ac.at/models.

International Monetary Fund, 2013, World Economic Outlook Database (Washington: International Monetary Fund). http://www.imf.org/external/pubs/ft/weo/2013/01/weodata/index.aspx.

Klimont, Zbigniew, Janusz Cofala, Imrich Bertok, Markus Amann, and Chris Heyes, 2002, "Modelling Particulate Emissions in Europe. A Framework to Estimate Reduction Potential and Control Costs" (Laxenburg, Austria: International Institute for Applied Systems Analysis).

Krewski, Daniel, Michael Jerrett, Richard T. Burnett, Renjun Ma, Edward Hughes, Yuanli Shi, Michelle C. Turner, C. Arden Pope III, George Thurston, Eugenia E. Calle, and Michael J. Thun, 2009, "Extended Follow-Up and Spatial Analysis of the American Cancer Society Study Linking Particulate Air Pollution and Mortality," Research Report 140 (Boston, MA Health Effects Institute). http://scientificintegrityinstitute.net/Krewski052108.pdf.

Krupnick, Alan, 2007, "Mortality-Risk Valuation and Age: Stated Preference Evidence," *Review of Environmental Economic Policy*, Vol. 1, pp. 261–82.

———, Anna Alberini, Maureen Cropper, and Nathalie Simon, 2000, "What Are Older People Willing to Pay to Reduce Their Risk of Dying?" Discussion paper (Washington: Resources for the Future).

Lepeule, J., F. Laden, D. Dockery, and J. Schwartz, 2012, "Chronic Exposure to Fine Particles and Mortality: An Extended Follow-up of the Harvard Six Cities Study from 1974 to 2009," *Environmental Health Perspectives*, Vol. 120, pp. 965–70.

Levy J., S. Wolff, and J. Evans, 2002, "A Regression-Based Approach for Estimating Primary and Secondary Particulate Matter Intake Fractions," *Risk Analysis*, Vol. 22, pp. 895–904.

Lükewille, A., and M. Viana, 2012, *Particulate Matter from Natural Sources and Related Reporting under the EU Air Quality Directive in 2008 and 2009*, EEA Technical Report 10/2012 (Copenhagen: European Environment Agency).

Muller, Nicholas Z., and Robert Mendelsohn, 2012, *Using Marginal Damages in Environmental Policy: A Study of Air Pollution in the United States* (Washington: American Enterprise Institute).

National Research Council, 2008, *Estimating Mortality Risk Reduction and Economic Benefits from Controlling Ozone Air Pollution* (Washington: National Academies Press).

———, 2009, *Hidden Costs of Energy: Unpriced Consequences of Energy Production and Use* (Washington: National Academies Press).

Nielsen, Chris P., and Mun S. Ho, eds., 2013, *Clearer Skies over China: Reconciling Air Quality, Climate, and Economic Goals* (Cambridge, Massachusetts: MIT Press).

Nordhaus, William D., 2013, *The Climate Casino: Risk, Uncertainty, and Economics for a Warming World* (New Haven, Connecticut: Yale University Press).

Organization for Economic Cooperation and Development (OECD), 2012, *Mortality Risk Valuation in Environment, Health and Transport Policies* (Paris: Organization for Economic Cooperation and Development).

Ostro, B., 2004, "Outdoor Air Pollution—Assessing the Environmental Burden of Disease at National and Local Levels," in *Environmental Burden of Disease Series*, edited by A.

Prüss-Üstün, D. Campbell-Lendrum, C. Corvalán, and A. Woodward (Geneva: World Health Organization).

Pindyck, Robert S., 2013, "Climate Change Policy: What Do the Models Tell Us?" *Journal of Economic Literature*, Vol. 51, No. 3, pp. 860–72.

Schaap, M., E.P. Weijers, D. Mooibroek, L. Nguyen, and R. Hoogerbrugge, 2010, *Composition and Origin of Particulate Matter in the Netherlands* (Bilthoven: PBL Netherlands Environmental Assessment Agency).

Stern, Nicholas, 2007, *The Economics of Climate Change* (Cambridge, U.K.: Cambridge University Press).

United Nations Environment Programme (UNEP), 2011, *Opportunities to Limit Near-Term Climate Change: An Integrated Assessment of Black Carbon and Tropospheric Ozone and Its Precursors* (Nairobi: United Nations Environment Programme and World Meteorological Organization).

United States Environmental Protection Agency (US EPA), 2011, *The Benefits and Costs of the Clean Air Act from 1990 to 2020*, Report to Congress (Washington: US Environmental Protection Agency).

United States Inter-Agency Working Group (US IAWG), 2013, *Technical Update of the Social Cost of Carbon for Regulatory Impact Analysis Under Executive Order 12866* (Washington: United States Government).

Weitzman, Martin L., 2009, "On Modeling and Interpreting the Economics of Catastrophic Climate Change," *Review of Economics and Statistics,* Vol. 91, No. 1, pp. 1–19.

World Bank, 2013, World Development Indicators Database (Washington: World Bank). http://data.worldbank.org/indicator.

World Bank and State Environmental Protection Agency of China, 2007, *Cost of Pollution in China: Economic Estimates of Physical Damages* (Washington: World Bank).

Zhou, Ying, Jonathan I. Levy, John S. Evans, and James K. Hammitt, 2006, "The Influence of Geographic Location on Population Exposure to Emissions from Power Plants throughout China," *Environment International*, Vol. 32, pp. 365–73.

CHAPTER 5

Measuring Nonpollution Externalities from Motor Vehicles

This chapter consists of three sections focused on the three major, non-pollution-related externalities from motor vehicles: traffic congestion, traffic accidents, and (to a much lesser extent) wear and tear on the road network (relevant for trucks). Other data and assumptions needed to implement the corrective motor fuel tax formulas from Chapter 3 are discussed in the annexes to this chapter.

CONGESTION COSTS

Basically, what is needed here is the cost of reduced travel speeds for other road users caused by an extra kilometer of driving by one vehicle, averaged across different roads in a country and across times of day. This cost estimate can then be used in the formula for corrective motor fuel taxes (equation (3.1)). As noted in Chapter 3, to manage congestion on the road network most effectively, countries should ideally transition to kilometer-based taxes that vary with the prevailing degree of congestion on different roads. However, until these schemes are comprehensively implemented, charging motorists for congestion costs through fuel taxes is entirely appropriate.

The congestion cost has two main components: First is the average added travel delay to other road users, as defined more technically in Annex 5.1, which (because of lack of direct data) needs to be extrapolated to the country level. Second, to convert delays into a monetary cost is the value of travel time (VOT), which is related to local wage rates.

The chapter begins by using a city-level database (covering numerous countries) to establish statistical relationships between congestion delays and various transportation indicators. These results, and country-level data for those same indicators, are then used to extrapolate congestion delays to the country level. Next, the way in which delays are converted into congestion costs is discussed. Results are presented, then a quick check on the results is performed by comparing them with cost estimates obtained from detailed country-level data (for a couple of countries for which these data are readily available).

The focus is on the most important cost component (time lost to motorists). Box 5.1 reviews some broader costs that should, in principle, be factored into corrective fuel tax assessments, but that are beyond the scope of this volume. For this reason, along with other assumptions made below, the congestion cost estimates in this chapter are probably on the low side.

BOX 5.1

Broader Costs of Congestion

One additional cost, beyond the pure time losses from travel delay, is the added fuel cost to other motorists from the possible deterioration in fuel efficiency experienced under congested conditions. The link between slower travel speeds and fuel consumption rates is complicated, however (Greenwood and Bennett, 1996; Small and Gómez-Ibáñez, 1998). Sometimes congestion slows traffic without increasing stop-and-go conditions, which could improve fuel efficiency for some range of relatively fast travel speeds. For the United States as a whole, Schrank, Lomax, and Eisele (2011, p. 5) estimate added fuel costs to be about 5 percent of the total costs of congestion, suggesting that these costs may have only modest implications for corrective fuel taxes.

Other, more subtle, costs of congestion may be more significant. For example, people may choose to set off earlier or later to avoid the peak of the rush hour, which may cause them to arrive earlier or later at their destination than they would otherwise prefer, perhaps because early arrival means they waste time waiting for an appointment, or late arrival runs the risk of penalties at work. Furthermore, congestion can result in day-to-day uncertainty about travel times, making it more difficult to plan the day (e.g., scheduling appointments, dinner times, and day care pickups). Studies suggest that travel time variability alone might raise the overall costs of congestion by about 10–30 percent (Eliasson, 2006; Fosgerau and others, 2008; Peer, Koopmans, and Verhoef, 2012).

Travel Delays at the City Level

The starting point is the Millennium Cities Database for Sustainable Transport, which provides detailed information on transportation in 100 cities (10 are discarded because of missing data).[1] These cities are listed in Annex 5.2.

The data are for 1995 and therefore rather dated, though they are the best available. Moreover, the age of the data need not be a problem because they are used for estimating statistical relationships between travel delay and transportation indictors, which are then matched with recent, country-level data on those indicators to provide up-to-date country-level estimates of travel delay. This approach is reasonable so long as the statistical relationships between delays and transportation indicators have not changed substantially since 1995.

The average road network speed in the database is the average speed of all motor vehicles (average for 24 hours/day, 7 days/week) on all classes of road in the metropolitan area.[2] These data provide information on recurrent congestion delays (occurring each day under normal driving conditions) but not on the average amount of nonrecurrent congestion (occurring from sporadic events such as accidents, bad weather, and road work). In this sense congestion costs are understated, perhaps significantly.[3]

---

[1] The database was developed by the International Association of Public Transport (UITP) and the Institute for Sustainability and Technology (ISTP) in 2001.
[2] Most of the speed data are calculated using traffic counts and assumptions about how speed varies with traffic volume.
[3] A study for Canada, for example, suggests that nonrecurrent congestion costs could be as large as those for recurrent congestion (Transport Canada, 2006).

TABLE 5.1

City-Level Travel Delays and Other Characteristics, Region Average, 1995

Region	Number of cities	Average speed (km/ hour)	Average delay (hours/ km)	Metropolitan GDP (1995 US$ per capita)	Annual km driven per car (thousands)	Road capacity (km/car)	Cars per capita
Africa	7	33.6	0.0159	2,500	11.8	33.2	0.10
Asian Affluent Cities	5	31.3	0.0164	34,800	12.2	16.3	0.22
Other Asian Cities	12	20.6	0.0342	4,200	10.5	20.0	0.09
Eastern Europe	5	31.3	0.0164	5,600	7.6	8.1	0.31
Western Europe	33	32.9	0.0144	31,900	11.3	12.4	0.41
Latin America	5	29.4	0.0195	5,400	10.1	16.0	0.19
North America	15	47.7	0.0058	27,900	18.5	17.3	0.57
Middle East	3	36.9	0.0153	7,700	14.9	12.7	0.19
Oceania	5	44.2	0.0074	19,800	12.9	22.4	0.58
All Cities	90	34.2	0.0158	21,000	12.4	16.6	0.34

Sources: Millennium Cities Database; and for average delay, authors' calculations.
Note: Figures are simple averages across urban centers in different regions. The average road network speed is the average speed of all vehicles (24 hours/day; 7 days/week) on all classes of road in the metropolitan area.

As indicated in Table 5.1, across all cities the average travel speed is 34.2 kilometers/hour, with speeds well above this average in North American cities (47.7 kilometers/hour) and well below it in non-affluent Asian cities such as Delhi (20.6 kilometers/hour).

The speed data are used to derive average travel delays using assumptions about travel speeds that would occur in the absence of congestion.[4] As indicated in Table 5.1, these estimated average delays per vehicle-kilometer are lowest in North America (0.006 hours/kilometer), and more than twice as large in western and eastern Europe, the Middle East, Africa, and affluent Asian cities such as Tokyo. Delays per kilometer are greater still in Latin American cities and non-affluent Asian cities.

Statistical regressions are used to obtain a relationship for predicting average delays for countries as a whole, using common indicators that are available both for the 90 cities in the Millennium Cities Database and in the country-level data discussed below. These variables comprise the following:

- Metropolitan GDP per capita (an indicator of a city's level of economic development)
- Annual car-kilometers (an indicator of traffic mobility)
- Road length or capacity per car
- Cars in use per capita (this, and the previous variable, are indicators of traffic intensity, relating to transport infrastructure and supply).

---

[4] These free flow speeds (which are not available in the data) are assumed to be 57 kilometers/hour (35 miles/hour) or 65 kilometers/hour (40 miles/hour), according to whether cities have relatively high or relatively low road density per urban hectare (in fact, for some cities, the observed travel speeds are close to the free flow speeds). These assumptions are roughly in line with those in Parry and Small (2009).

Common statistical techniques are used to estimate coefficients that show the contribution of each of these indicators toward explaining average travel delays across cities, using functional forms that best fit the data. Further details, along with the statistical regression results, are provided in Annex 5.3.

Ideally, additional variables would be included in these regressions to improve statistical accuracy. However, because the purpose is to make country-level extrapolations, only those indicators for which data are available at the country level can be used. Despite this limitation, a reasonably good statistical fit is still obtained.

Projecting Country-Level Delays

The estimated statistical relationships between the average delay and the four key indictors at the city level are now used to project the average delay for 150 countries (all countries for which these data are available), with the country-level indicators. For this purpose, GDP per capita is taken from World Bank (2013) and all other indicators from *World Road Statistics 2009* (IRF, 2009).[5] For 81 of the countries, data on car-kilometers traveled are missing. Annex 5.3 describes how this data gap was filled using supplementary statistical regressions.

Table 5.2 summarizes the key indicators by region. At the country level, per capita incomes are lower, annual kilometers driven per car are higher, and road capacity per car is smaller than in the city-level data shown in Table 5.1.

The estimated coefficients from the city-level analysis are used together with country-level variables to predict the average nationwide delays in the 150 countries. Because the city-level regression is based on 90 major cities, the average predicted delay represents the urban congestion level for each country, excluding the rural areas. To predict the average delay at the country level, the predicted

TABLE 5.2

Country-Level Travel Delays and Other Characteristics, Region Average, 2007

Region	Number of countries	Predicted average delay (hours/km)	Country GDP (2007 US$ per capita)	Annual km driven per car (thousands)	Road capacity (km/car)	Cars per capita
Africa	45	0.0046	2,300	36.3	1,395	0.03
Asia	33	0.0053	9,900	16.3	362	0.11
Europe	43	0.0025	26,900	9.4	65	0.35
Latin America	11	0.0049	5,100	21.9	185	0.09
North America	11	0.0048	12,100	19.5	103	0.15
Oceania	7	0.0028	11,900	18.1	290	0.20
All Countries	150	0.0041	12,400	21.0	551	0.16

Sources: IRF (2009); and authors' estimates of some data on annual km driven per car (see Annex 5.3). Average delay is predicted using procedures described in the text.
Note: km = kilometer. Amounts are simple averages across countries—therefore, for example, high average delays in Mexico inflate the average amount for North American countries.

---

[5]The most recent data are for 2007, which is assumed to provide a reasonable approximation for delays in 2010.

urban average delay is scaled by the urban population ratio, on the assumption that rural congestion is negligible.

Comparing the results in the second column of Table 5.2 with those from Table 5.1, the average vehicle delays at the country level are about one-quarter to one-half of those at the city level. This difference makes sense—the city level data focus only on delays in large cities (where congestion is especially severe), whereas the country-level estimates also account for driving in rural areas and medium and small cities. Nonetheless, it is important to bear in mind that average travel delays at the country level are estimated with a fair amount of imprecision, especially for countries for which there might be substantial errors in the measurement of transportation indicators.[6]

From Delays to Congestion Costs

This section explains how delays that one vehicle imposes on others are derived from the above estimates and then monetized. Complications posed by other vehicles on the road, such as buses, are also discussed.

Deriving delays imposed on others by one vehicle

A specification commonly used by transportation engineers for the relationship between travel speed or time and traffic volume results in a simple relationship between average delays per kilometer (estimated above) experienced by individual drivers and the increased travel time that one extra vehicle implies for all other vehicles on the road.

When travel delay is a simple power function of traffic volume relative to road capacity, with the exponent in this function denoted by β, then the extra delay one vehicle imposes on other vehicles is simply β times the average delay per kilometer (see Annex 5.1). Empirical studies suggest that β is roughly in the range of 2.5–5.0, with higher values in this range applicable to larger urban centers. In this analysis, β is assumed to be 4.[7]

Finally, delays to other passengers are obtained by multiplying delays to other vehicles by the vehicle occupancy rate, assumed to be 1.6 (Annex 5.4).[8] Alternative assumptions about vehicle occupancy and the exponent β would have

---

[6] In a handful of cases for which the results looked especially questionable, average delays per kilometer were extrapolated from other countries. For example, delays for Bangladesh and Kazakhstan were extrapolated from India and Russia, respectively, and included an adjustment for differences in urbanization rates between countries.

[7] This assumption is consistent with the Bureau of Public Roads formula, the traditional method for predicting vehicle speed as a function of the volume-to-capacity ratio. See Small and Verhoef (2007, pp. 69–83) and Small (1992, pp. 70–71) for further discussion. Obviously the above approach is highly simplified—speed/volume relationships may vary considerably with the characteristics of specific roads (e.g., speed limits, frequency of stop lights and sharp bends) and across different times of day. But the assumed value for β seems to be a reasonable rule of thumb for representing average travel conditions in urban areas.

[8] This is slightly higher than the average vehicle occupancy rates for London, Los Angeles, and Washington calculated in Parry and Small (2009).

proportional effects on the congestion costs reported below (e.g., if β = 5 or average vehicle occupancy is 2, congestion costs would be 25 percent greater).

Value of travel time (VOT)

The discussion now turns to the VOT, which is needed to monetize congestion costs.

According to economic theory (Becker, 1965), on average, people should organize their time such that they are indifferent between an extra hour at work and an extra hour of nonmarket time (e.g., relaxing at home, looking after the children). Therefore, an extra hour of nonmarket time is commonly valued by the benefit to individuals of an extra hour of forgone work, namely, the after-tax hourly wage (i.e., the market wage after netting out personal income and employee payroll taxes, and consumption taxes paid when wages are spent).

As a first pass, people might also value an extra hour of travel time by the net-of-tax wage, which would suggest a VOT of about 50–70 percent of the market wage for a typical advanced country. More generally, the monetary cost of travel time could be lower (if people enjoy driving, for example, because they can listen to music) or higher (if people enjoy the workplace, for example, because of interaction with colleagues).

A large empirical literature estimates the VOT for personal travel using revealed and stated preference techniques similar to those discussed in Chapter 4. A revealed preference study might involve, for example, estimating people's willingness to pay extra auto fuel and parking costs to save time as compared with an alternative, slower travel mode, while a stated preference study might involve directly asking people what tolls they might pay for a faster commute.

For Canada, France, the United Kingdom, and the United States, literature reviews suggest that a VOT of about half the market wage is a reasonable rule of thumb for general automobile travel (see Table 5.3). The VOT is somewhat higher for commuting (e.g., because of penalties for late arrival at work) than for non-market-related trips such as shopping, taking the children to school, or going to the gym—16 percent higher according to Wardman (2001). Here the VOT is assumed to be 60 percent of the market wage, given that most delays occur during the commuter-dominated peak period.

The VOT-to-market wage ratio is assumed to be the same across all countries.[9] The wage data are from the International Labor Organization's Global Wage Database (ILO, 2012) and are nationwide measures for 2010.[10]

---

[9]Much evidence, at least from advanced countries, suggests that the VOT increases approximately in proportion to income, which backs up this assumption (Small and Verhoef, 2007, p. 52). Abrantes and Wardman (2011) determine that a 10 percent increase in income increases the VOT by 9 percent. It might be argued that the VOT should be adjusted upward in countries with relatively low vehicle ownership rates, where ownership is skewed toward higher wage groups. No adjustments are made, however, partly because of data limitations. But the issue is not clear cut either—conceivably, higher-income people (at least those living in more expensive housing closer to downtown areas) drive less under congested conditions than do other motorists.

[10]There are data gaps for six countries in ILO (2012). For these cases, wages are proxied using GDP per capita. Ideally, urban wage rates (adjusted downward for differences compensating for higher

TABLE 5.3
Reviews of Empirical Literature on the Value of Travel Time (VOT)

Study	About the study	Recommended VOT (percent of market wage)
Waters (1996)	Reviews 56 estimates from 14 countries	35–50
Wardman (1998)	Review of U.K. studies	52
Mackie and others (2003)	Review of U.K. studies	51
US Department of Transportation (1997)	Review of U.S. studies	50
Transport Canada (1994)	Review of U.S. and Canadian studies	50
Commissariat General du Plan (2001)	Review of French studies	59

Note: Summary findings for these reviews were taken from Small and Verhoef (2007, pp. 52–53). Studies take a weighted average over different trip types (usually at peak period) except for Waters (1996) who focuses exclusively on commuter trips.

Figure 5.1 shows the VOT for selected countries. Broadly speaking, the relative pattern of VOTs across countries is similar to that for the value of mortality risks in Figure 4.2 of Chapter 4.[11]

Accounting for other vehicles

The estimates in this analysis assume that all vehicles on the road are cars, whereas in practice the vehicle fleet comprises a mixture of cars, buses, trucks, and two-wheel motorized vehicles. Annex 5.4 discusses and applies a formula that shows the ratio of congestion costs (properly estimated accounting for the mix of vehicles) relative to the congestion cost estimated here.

If trucks and two-wheelers account for a sizable portion of the vehicle fleet (but buses do not) the estimates are not very different. However, if buses account for a significant portion of vehicle-kilometers, the estimates here can substantially understate congestion costs (see the Annex 5.4). Cars have a significantly greater impact on increasing travel times for other road users when a greater portion of vehicles on the road are carrying large numbers of passengers. An adjustment is not made here, however, because data on the share of buses in urban vehicle-kilometers are not available for many countries.[12]

---

living costs) would be used in preference to nationwide wages, but a comprehensive, international data set is not available.

[11]There are some nuances. The relative differences between developed and developing countries are a bit more pronounced for the VOT because relative wages across countries are compared whereas Figure 4.2 compares relative income raised to the power 0.8. There are also some differences even among similar-income countries. For example, the United States has a higher mortality valuation than Australia but slightly lower VOT, reflecting the depressing effect on U.S. wages of relatively high labor force participation among migrants and secondary family workers, and relatively little influence of labor unions or labor market regulations on inflating wages.

[12]In many cases the bus share is very low (e.g., about 1 percent or less of vehicle-kilometers traveled in Washington, Los Angeles, and London—see Parry and Small, 2009).

Figure 5.1 Value of Travel Time, Selected Countries, 2010

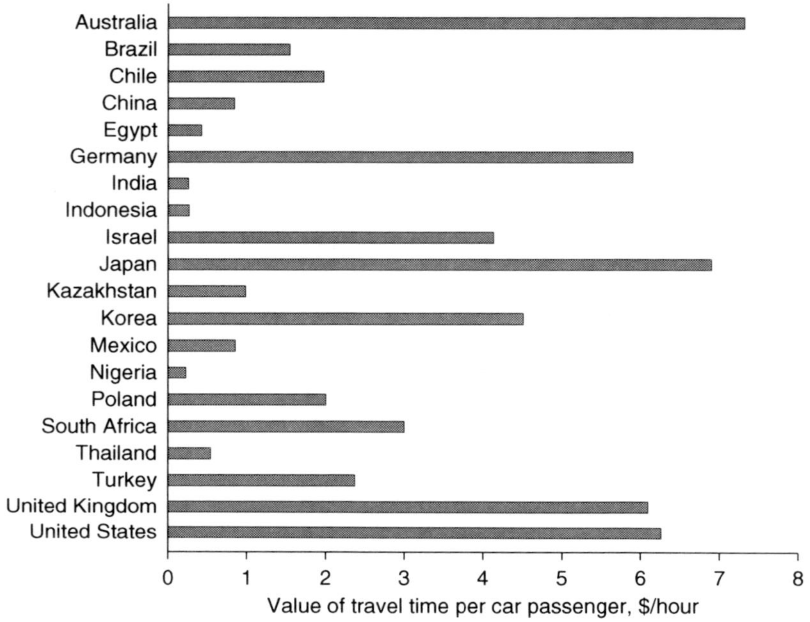

Country	Value of travel time per car passenger, $/hour
Australia	~7.4
Brazil	~1.5
Chile	~2.0
China	~0.9
Egypt	~0.4
Germany	~5.9
India	~0.2
Indonesia	~0.2
Israel	~4.1
Japan	~6.9
Kazakhstan	~1.0
Korea	~4.5
Mexico	~0.9
Nigeria	~0.2
Poland	~2.0
South Africa	~3.0
Thailand	~0.5
Turkey	~2.4
United Kingdom	~6.1
United States	~6.3

Source: Authors' calculations.

Finally, in the computation of corrective diesel fuel taxes, an extra truck-kilometer is assumed, based on the literature (Lindsey, 2010, p. 363; Transportation Research Board, 2010; Parry and Small, 2009), to contribute twice as much to congestion as an extra car-kilometer (trucks drive more slowly and take up more road space, though a partially offsetting factor is that they tend to be driven less intensively on congested roads).

Results

Figure 5.2 shows nationwide congestion costs imposed on others per extra car-kilometer, for 20 selected countries.

The congestion cost for the United Kingdom, for example, is US$0.09/kilometer. Australia, Germany, Israel, Korea, and South Africa all have broadly similar congestion costs, while Turkey's is substantially higher, and Japan's higher still. (Although Japan has a relatively high VOT, most of the difference is due to its greater estimated travel delays.) Congestion costs for the United States, where a smaller portion of nationwide driving occurs under congested conditions, are lower at US$0.064/kilometer (though this U.S. estimate seems on the high side relative to a potentially more accurate estimate discussed below). China's estimated congestion cost is US$0.05/kilometer, less than the United States—despite China having greater average travel delays—because of its much lower assumed VOT. Low VOTs also help explain the low congestion costs (less than US$0.01/kilometer) in India and Kazakhstan.

Figure 5.2 Congestion Costs Imposed on Others per Car-Kilometer, Selected Countries, 2010

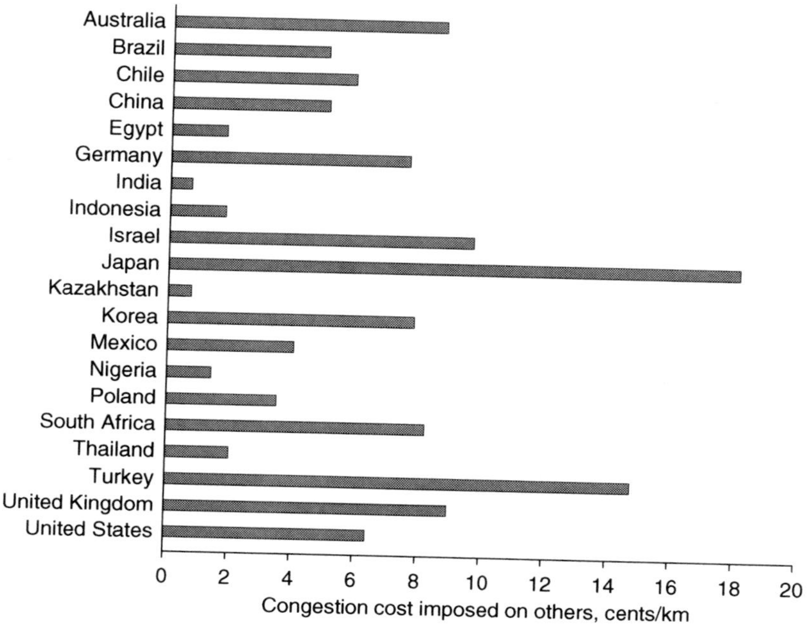

[Bar chart showing congestion cost imposed on others, cents/km, for the following countries (approximate values):
- Australia: ~9
- Brazil: ~5
- Chile: ~6
- China: ~5
- Egypt: ~2
- Germany: ~8
- India: ~1
- Indonesia: ~2
- Israel: ~10
- Japan: ~18
- Kazakhstan: ~1
- Korea: ~8
- Mexico: ~4
- Nigeria: ~2
- Poland: ~4
- South Africa: ~8
- Thailand: ~2
- Turkey: ~14
- United Kingdom: ~9
- United States: ~6]

Source: Authors' calculations.

Figure 5.3 shows ranges of estimated congestion costs for all countries, where data allow. Again, these costs are relatively high in western Europe (where a large portion of driving occurs under congested conditions and people have a high VOT) and, except for South Africa, relatively low in Africa (where the VOT is lowest). The United States, Latin America, and Australia are intermediate cases.

Robustness Checks

For the United Kingdom and the United States, detailed data on travel delays for road classes in different regions are available and can be combined into an alternative estimate of nationwide average delay as a check on the above estimates (see Annex 5.5 for estimation procedures and data sources).

For the United Kingdom, the average delay per vehicle-kilometer obtained from this alternative data source is almost exactly (within 1 percent of) that estimated above, providing some reassurance that the approach, at least for the United Kingdom, might be reasonable. For the United States, the average delay using the alternative data is 59 percent of that estimated above, suggesting that the estimate in this analysis may be on the high side for that particular country.

The delay estimates from country-level data should be more reliable than the extrapolations presented above, though the former are surprisingly hard to come by (transportation authorities do not routinely collect these data). The

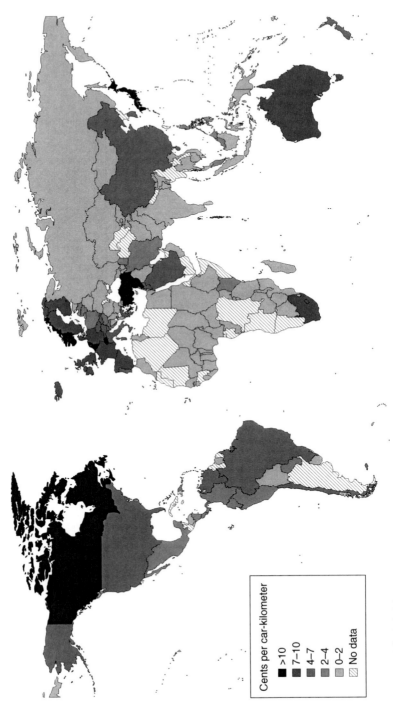

Figure 5.3 Congestion Costs Imposed on Others per Car-Kilometer, All Countries, 2010

Source: Authors' calculations.

approach used in this chapter suffers from imprecision given the limited number of indicators common to both the city- and country-level data, issues with the quality of both city- and country-level data sets, and the possibility that the underlying relationship between the average delay and city or country characteristics might have evolved with changes in infrastructure, technology, and traffic rules. However, it is hard to gauge the direction, let alone the magnitude, of bias for individual countries. Moreover, broadly speaking the pattern of relative congestion costs across different countries estimated above seems plausible, even though the individual country estimates may not be especially accurate.

ACCIDENT COSTS

The total societal costs from road traffic accidents can be substantial, and are often underappreciated. Gauging the appropriate charge for accident risk to be reflected in fuel taxes as a complement to other measures such as road safety investments is difficult, however, for two reasons.

First, conceptually it is a bit tricky to judge whether certain categories of costs should be viewed as "internal" because individuals take them into account in their driving decisions, or "external," that is, borne by others. (Only the latter warrant corrective taxes.)

Second, although data are often available for road fatalities, they are usually not available for other accident costs such as nonfatal injuries, medical and property damage costs, and even fatality data are not always broken down in a way that permits assessment of external costs.

The estimates in this chapter necessarily rely on some judgment calls, extrapolations to fill in missing costs, and transfers of fatality breakdowns across similar countries. For these and other reasons, the accuracy of cost estimates can be questioned. But again, the estimates provide some plausible and transparent sense of external accident costs, shed light on why these costs differ across countries, and highlight the data needed to improve the future accuracy of cost assessments.

The discussion proceeds as follows: conceptual issues are reviewed in an attempt to categorize different accident costs into internal versus external risks; the estimation of external costs is discussed; then results are presented.

Classifying Accident Risks: Some General Principles

The main societal costs of road accidents include personal costs of fatal and nonfatal injuries, medical costs, and property damage.[13]

---

[13]Other costs from traffic accidents, such as those from traffic holdups, police and fire services, insurance administration, and legal costs, are beyond the scope of this chapter. According to some studies, they appear to be modest relative to other costs (e.g., US FHWA, 2005; Parry, 2004, Table 2). That might seem surprising for traffic holdups given that some accidents cause severe traffic disruptions, but these accidents constitute only a small share of total accidents. Productivity losses are taken into account in the monetary values assigned to different types of injuries.

Injuries

Injury risks to pedestrians and cyclists, to vehicle occupants in accidents involving only a single vehicle, and to vehicle occupants in accidents involving multiple vehicles are considered separately.

Pedestrian and cyclist injuries: It is normally assumed that motorists do not take into account injury risks they pose to pedestrians and cyclists when deciding how much to drive (Newbery, 1990; Parry, 2004).[14] Such risks are therefore classified as external.

Injury risk to occupants in single-vehicle collisions: For accidents involving one vehicle, it is standard to view the injuries to occupants of such vehicles as risks that are taken into account: in other words, if individuals put themselves at greater risk (by getting in the car more often), this is not viewed as a basis for taxation to deter this behavior.[15] For similar reasons, injury risks to other occupants (e.g., family members) in single-vehicle collisions are generally viewed as internalized risks.

Injury risks to occupants in multivehicle collisions: Here the delineation between internal and external risks becomes murky. The issue is how extra driving by one vehicle affects injury risks to occupants of other vehicles. All else the same, extra driving by one motorist leads to more cars on the road and greater risks to others—cars have less road space on average and are therefore more likely to collide. In this case, injury risks to other vehicle occupants would increase approximately in proportion to the amount of traffic.

However, all else might not be the same: with more vehicles on the road, motorists may drive more carefully or be obliged to drive more slowly. Thus, an offsetting reduction in accident frequency and in the average severity of injuries in a given accident (because vehicles collide at slower speeds) might occur. Although slightly slower driving may not do much to reduce injury risks to unprotected pedestrians, the effect may be more pronounced for other vehicle occupants, who have a greater degree of protection. What matters, then, is the impact of additional driving on the "severity-adjusted" injury risk to other vehicle occupants. However, available evidence is inconclusive.[16]

An intermediate assumption between the two more extreme cases is considered in this analysis. In one case, additional driving leads to a proportionate increase

---

[14]However, once on the road, drivers likely take care to lower the risks of hitting pedestrians and cyclists. Because observed injury data reflect this likelihood, it is taken into account in the corrective tax estimates.

[15]Motorists may lack an accurate sense of risks to themselves but, in the absence of evidence to the contrary, it seems reasonable to assume that the average motorist does not systematically understate or overstate these risks—and even if there were such evidence, information campaigns to better educate drivers might be a better response than corrective fuel taxes.

[16]For example, Edlin and Karaca-Mandic (2006) find that extra driving substantially increases average insurance costs per kilometer driven, suggesting higher per kilometer property damage costs (though how other costs, like fatality risk, change is not clear). However, studies by Lindberg (2001), Traynor (1994), and Fridstrøm and others (1995) suggest that extra driving may have only limited effects, and possibly even a negative effect, on severity-adjusted accident risk.

in injury risks to others (there is no offsetting decline in severity-adjusted accident risk due to slower or more careful driving). In the second case, extra driving has no effect on severity-adjusted injury risks to others; increased risk to others is completely offset by a decline in the average severity of injuries.

In the first case, it is assumed that half of injuries in multivehicle collisions are external based approximately on the logic that, on average, one vehicle is responsible for the collision and another is not and that those at fault take into account risks to occupants of their vehicles but not occupants of other vehicles (Parry, 2004). In the second case, all injuries in multivehicle collisions are internal. Splitting the difference suggests that one-quarter of multivehicle collision injuries should be treated as external.

Medical and property damage costs

Medical costs associated with all traffic-related injuries are largely borne by third parties (the government or insurance companies), though individuals typically bear some minor portion of these costs through, for example, copayments and deductibles.

It is difficult to pin down how much property damage, primarily repairs or replacement costs for damaged vehicles, drivers take into account. In countries with comprehensive insurance systems, some costs are borne by third parties (insurance companies) but other costs are borne by drivers in the form of deductibles and possibly elevated future premiums following a crash.[17]

Accident risks from heavy vehicles

Accident costs for trucks are needed to compute the corrective diesel fuel tax. At first glance, it might appear that trucks would impose much greater risks to other road users than cars, given their much greater weight. A counteracting factor, however, is that trucks are driven at slower speeds than cars and that truck drivers are professionals, which may further reduce their crash risk (e.g., because truck drivers are unlikely to drink and drive).[18] According to a detailed study by the United States Federal Highway Administration (US FHWA, 1997, Table V-24), overall external accident costs per vehicle-kilometer are only slightly higher for heavy vehicles than for cars—therefore, these costs are assumed in this analysis to be the same for cars and trucks (see also Parry and Small, 2009).

External Cost Assessment

IRF (2012) provides data on traffic fatalities for 2010 or the latest available for most countries, based on local data, such as from police reports. WHO (2013)

---

[17] Premiums may also vary moderately with an individual's stated annual driving, which also provides some, albeit very weak, link between extra driving and property damage (in the form of greater premiums) paid by drivers. And to the extent that insurance companies have market power, motorists may be taxed, in effect, for risks of property damage.

[18] In 2010, the crash frequency per kilometer driven in the United States for light-duty vehicles was almost four times that for trucks (BTS, 2012, Tables 2.21 and 2.23).

provides data on the breakdown of fatalities by pedestrians, cyclists, occupants of motorized two- to three-wheelers, occupants of four-wheelers, and a miscellaneous category (e.g., bus riders). In cases in which only total fatalities are reported, the breakdown is assumed to be the same as in another, similar country in the same region. The data used here likely underreport, perhaps substantially, road fatalities for many developing countries, providing yet another reason the corrective fuel tax estimates presented later might be understated.[19]

The vehicle occupant data do not separate out deaths in multivehicle collisions from those in single-vehicle collisions. Based on a simple average across five country case studies (discussed in Annex 5.6), 57 percent of fatalities of occupants of two-, three-, and four-wheelers are assumed to occur in multivehicle collisions. And from the previous discussion, 25 percent of these are external fatality risks, as are all of the pedestrian and cyclist fatalities. The same values by country as used in Chapter 4 for pollution deaths are used to monetize these fatalities.[20]

Data are not available for most countries for other components of external accident costs—nonfatal injuries, medical costs, and property damage. However, based on comprehensive estimates of these costs for Chile, Finland, Sweden, the United Kingdom, and the United States, a relationship between the ratio of these other external costs[21] to the external costs of fatalities was estimated as a function of the share of external fatalities in total fatalities (Annex 5.6); in countries with a high incidence of pedestrian deaths, the relative size of other external costs tends to be smaller. The external cost ratio was then inferred for different countries based on their shares of external fatalities in total fatalities, and the external costs were scaled up accordingly.

Results

Figures 5.4 and 5.5 show, respectively, the external accident costs for selected countries and for all countries expressed, to facilitate comparison with congestion costs per vehicle-kilometer of travel by car or truck. (See Annex 5.3 on measurement of vehicle-kilometers.)

Higher-income countries tend to have a lower incidence of injuries per kilometer driven because as countries develop, vehicle and road safety tend to improve, and the ratio of pedestrians and cyclists to motorists declines (Kopits and Cropper, 2008).[22] This lower incidence of injuries is partially, but not entirely, offset by higher

---

[19] For example, fatalities in India were 133,938 for 2010 according to IRF (2012), but were estimated to be 231,027 in 2010 by WHO (2013).

[20] In principle, it might seem that a higher value should be used for traffic-related deaths, given that the average age of someone dying in a road accident is lower than for the average person dying from pollution-related illness (Small and Verhoef, 2007, p. 101). However, for reasons discussed in Box 4.3 of Chapter 4, an adjustment is not made.

[21] Property damage accounts for 42 percent of other external costs, nonfatal injuries 38 percent, and medical costs 20 percent, based on a simple average across the studies.

[22] In India, for example, there are 40 external deaths per billion vehicle-kilometers compared with 2 in the United States.

Figure 5.4 External Accident Costs per Vehicle-Kilometer, Selected Countries, 2010

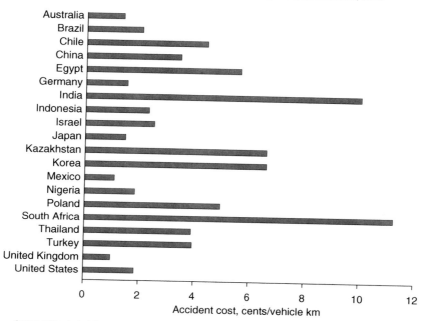

Source: Authors' calculations.
Note: Figure shows external accident costs (reflecting fatal and nonfatal injuries, medical costs, and property damage) expressed per kilometer driven by cars or trucks.

valuations of fatality and injury risk in higher-income countries. Loosely speaking, therefore, these figures show a pattern, with some exceptions, of lower external accident costs per kilometer in higher-income countries. For example, costs are significantly less than US$0.04/kilometer in Australia, Japan, western European countries, and the United States, and more than US$0.06/kilometer in some Central and South American and African countries, and in India, Kazakhstan, and Russia.

Note also (comparing Figures 5.2 and 5.3) that external accident costs can be of the same broad order of magnitude as congestion costs. In fact, in 11 of the selected countries, accident costs are greater than congestion costs.

ROAD DAMAGE COSTS

Vehicle use causes an additional adverse side effect through wear and tear on the road network. However, given that road damage is a rapidly rising function of a vehicle's axle weight, nearly all of the damage is attributable to heavy-duty vehicles; road damage costs for light-duty vehicles make little difference for corrective fuel taxes (US FHWA, 1997) and are ignored in this analysis.[23] Road damage

---

[23] Road damage increases approximately in proportion to the third power of a vehicle's axle weight (Small, Winston, and Evans, 1989), though there is considerable variation across road surfaces.

116 Getting Energy Prices Right: From Principle to Practice

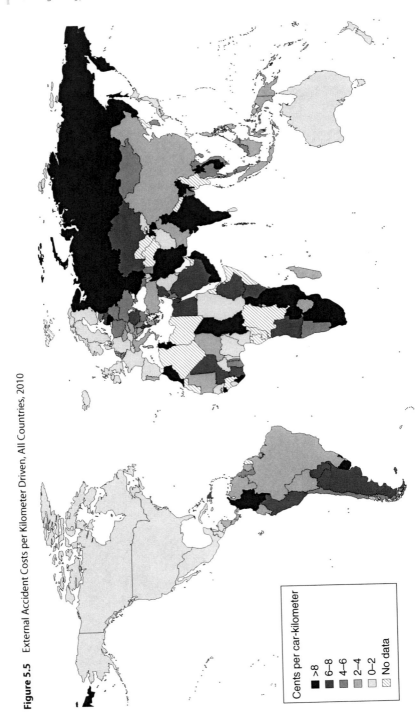

Figure 5.5 External Accident Costs per Kilometer Driven, All Countries, 2010

Source: Authors' calculations.

costs also vary considerably across different classes of trucks, which would matter for the design of a finely tuned system of axle weight tolls. However, the concern in this analysis is with damage caused by trucks as a group, so that it can be factored into the corrective diesel fuel tax.

Road damage consists both of the pavement repair costs incurred by the government and increased operating costs for vehicles attributable to bumpier roads. However, if the government steps in to repair roads once they reach a predetermined state of deterioration, then as a rough rule of thumb, the total external cost of road damage can be measured by average annual spending on maintaining the road network.[24]

A complication is that road damage is jointly caused by vehicle traffic and weather, such as ice creating and exacerbating holes and cracks in the pavement, which are further enlarged by vehicle traffic. Empirical work for apportioning damage to trucks versus weather is sparse. Depending on road strength (e.g., thickness), Paterson (1987, p. 372) suggests that vehicles cause 40–90 percent of damage in warm, dry, or subhumid climates; 20–80 percent in arid, freezing climates; and 10–60 percent in moist, freezing climates.[25] Here trucks are assumed to account for 50 percent of the damage in all countries.[26]

IRF (2009, Table 8.2) provides separately spending on road maintenance and capacity investments aggregated over all levels of government for 2007 (or the latest available) for 74 countries. For other countries, IRF (2009) provides total highway spending, but not the maintenance-to-investment decomposition; therefore, this breakdown is inferred from a similar country in the same region for which the breakdown is available. For the remaining 10 countries for which no spending data are available, maintenance expenditure per truck-kilometer is assumed to be the same as a similar country in the same region. Scaling by the share of trucks (as opposed to climate) in total damage gives the damage attributable to trucks by country.

SUMMARY

Congestion costs are estimated using extrapolations of travel delays from a city-level database to the country level (in the absence of direct data on these delays) and travel time valuations from the literature. Accident costs are assessed by making assumptions about what portion of road fatalities in different countries reflects risks that motorists do not take into account, and making some upward adjustments to allow for other components of accident risk (property damage, medical costs, and nonfatal injuries). Both congestion and accident costs are sizable (and

---

[24]If roads are repaired more frequently, government resource costs are higher but there is less deterioration of vehicle operating costs, and vice versa. See Newbery (1987) for a more precise discussion.
[25]See also Newbery (1987) for similar findings.
[26]A more accurate calculation (but one that would have little impact on overall corrective diesel fuel tax estimates) would involve classifying countries by climate zone and, better still, average road strength, and applying different assumptions about the proportion of damage attributable to trucks.

likely understated); in some cases, accident costs exceed congestion costs. Road damage is also estimated by attributing a portion of road maintenance expenditures to trucks, though these costs are modest in relative terms. A few extra steps are needed to calculate corrective motor fuel taxes; see Annex 5.7 for details.

All of the cost estimates are rudimentary and there will be ample scope for reforming them in the future as data, such as on travel delays, become more widely available and analytical work helps to resolve some of the uncertainties (e.g., about the VOT in low-income countries, or the safety risks that one driver poses to other road users). In the meantime, however, these cost estimates enable a first-pass estimate of corrective motor fuel taxes that can be studied and might serve as a useful starting point for discussions about tax reform.

ANNEX 5.1. MEASURING CONGESTION COSTS: SOME TECHNICALITIES

The total hourly costs (TC) of congestion to passengers in vehicles driving along a one-kilometer lane-segment of a highway can be expressed as

$$TC = V \times (T[V] - T^f) \times o \times VOT, \tag{5.1}$$

in which V denotes traffic volume or flow—the number of cars that pass along the kilometer-long stretch per hour (the implications of other vehicles on the road are discussed below). T^f is travel time per kilometer when traffic is freely flowing, and T (which exceeds T^f) is the actual travel time, an increasing function of the traffic volume (speeds fall with less road space between vehicles). The component o is vehicle occupancy, or average number of passengers per vehicle. The total travel delay from congestion for all passengers is therefore $V \times (T - T^f) \times o$, where $T - T^f$ is the average delay per vehicle-kilometer. Multiplying total travel delay by the value of travel time (VOT) expresses delays as a monetary cost.

Dividing TC by traffic volume gives the average cost of congestion (AC) per vehicle-kilometer:

$$AC = (T - T^f) \times o \times VOT. \tag{5.2}$$

AC is the cost borne by individual motorists that, on average, they should take into account when deciding how much to drive.

Differentiating TC with respect to V gives the added congestion cost to all road users from an extra vehicle-kilometer:

$$\frac{dTC}{dV} = AC + \frac{dT}{dV} \times V \times o \times VOT. \tag{5.3}$$

This added cost includes the average cost (taken into account by the driver), as just described. It also includes the cost to occupants of other vehicles, which is not taken into account by the driver. The latter is the delay to other vehicles, $(dT/dV) \times V$, times the average number of people in other vehicles, times the VOT to express costs in monetary units.

Suppose, as discussed in the main text, that travel delay can be approximated by a power function of traffic volume, that is,

$$T - T^f = \alpha V^\beta, \quad (5.4)$$

in which α and β are constants. The constant α reflects factors like road capacity, and β reflects the rate at which additional traffic diminishes travel speeds. Differentiating this expression by V gives $dT/dV = \alpha\beta V^{\beta-1}$. Then using (5.4) gives the following:

$$\frac{dT}{dV} \times V = \left(T - T^f\right) \times \beta. \quad (5.5)$$

The delay to other vehicles, $(dT/dV) \times V$, simply the product of average delay and the scalar β. As discussed in the main text, empirical studies suggest a value for β of between about 2.5 and 5.0 for congested roads.

If speed data are available, average delay can be estimated using the following equation:

$$T = \frac{1}{S}, \quad T^f = \frac{1}{S^f}, \quad (5.6)$$

in which S and S^f are the actual and the free-flow travel speeds (kilometers/hour).

ANNEX 5.2. CITIES COMPRISING CITY-LEVEL DATABASE

Cities covered in the city-level database, which is used to obtain statistical relationships between travel delays and various transportation indicators, are listed in Annex Table 5.2.1.

ANNEX TABLE 5.2.1

Cities in the City-Level Database (Used to Extrapolate Congestion Costs)

Western Europe	Eastern Europe	Middle East	Oceania
Amsterdam	Budapest	Riyadh	Brisbane
Athens	Cracow	Tel Aviv	Melbourne
Barcelona	Moscow	Tehran	Perth
Berne	Prague		Sydney
Berlin	Warsaw	**Africa**	Wellington
Bilogna		Abijan	
Brussels	**North America**	Cairo	
Copenhagen	Atlanta	Cape Town	
Dusseldorf	Calgary	Dakar	
Frankfurt	Chicago	Harare	
Geneva	Denver	Johannesburg	
Glasgow	Houston	Tunis	
Graz	Los Angeles		
Hamburg	Montreal	**Asian Affluent**	
Helsinki	New York	Hong Kong	

(Continued)

ANNEX TABLE 5.2.1 (*Continued*)

Cities in the City-Level Database (Used to Extrapolate Congestion Costs)		
Lille	Ottwa	Osaka
London	Phoenix	Sapporo
Lyon	San Diego	Singapore
Madrid	San Francisco	Tokyo
Manchester	Toronto	
Marseille	Vancouver	**Other Asian**
Milan	Washington	Bangkok
Munich		Beijing
Nantes	**Latin America**	Chennai
Newcastle	Bogota	Guangzhou
Oslo	Curitiba	Ho Chi Minh City
Paris	Mexico City	Jakarta
Rome	Rio de Janeiro	Kula Lumpur
Ruhr	San Paulo	Manila
Stockholm		Mumbai
Stuttgart		Seoul
Vienna		Shanghai
Zurich		Taipei

Source: See main text.
Note: Excludes 10 cities from the original database that were dropped because of insufficient data.

ANNEX 5.3. RESULTS FROM STATISTICAL METHODS USED TO RELATE TRAVEL DELAY TO TRAVEL INDICATORS

As discussed in the main text, statistical regressions were used to estimate the contribution of various factors to explaining travel delays across the 90 cities in the database. To obtain the best statistical fit (i.e., to reduce the noise from outlying observations or extreme values), average delay and the four explanatory indicators are expressed in natural logarithm form in the regression and second powers of these variables are included. (Both of these are standard statistical procedures.) The regression results are presented in Annex Table 5.3.1.

Interpreting these coefficients is less of a concern than the statistical fit (which is reasonably good) because the coefficients are used for prediction rather than for establishing causal relationships. In fact, the explanatory variables such as road length per car and cars per capita are likely to be simultaneously determined with traffic conditions such as average delay (the dependent variable), which confounds the interpretation of the coefficients.[27]

As noted in the main text, car-kilometers driven is not available at the country level for 81 countries. To fill in this gap, countries are grouped by region (Europe, Oceania, Africa, and so on) and statistical regressions are used to estimate a relationship for each regional grouping between car-kilometers (for countries for

---

[27] For example, the negative sign for kilometers per car suggests, perhaps, that extra traffic creates pressure or incentives for road investment (Duranton and Turner, 2011) or that bad traffic conditions discourage driving.

ANNEX TABLE 5.3.1
Regression Results for City-Level Average Delay

Variables	Log average delay
log GDP per capita	0.061
	−0.409
log km driven per car	−5.308***
	−1.776
log road length per car	−0.796
	−1.08
log cars per capita	−1.038***
	−0.242
log GDP per capita²	−0.0106
	−0.044
log km driven per car²	−0.515**
	−0.196
log road length per car²	−0.0414
	−0.11
log cars per capita²	−0.100*
	−0.051
Constant	−21.23***
	−5.04
Observations	90
R-squared	0.659

Source: See main text.
Note: *, **, and *** indicate significance at the 10 percent, 5 percent, and 1 percent levels, respectively.

which these data are available) and four explanatory variables that are available for all countries: per capita income, urban population density, vehicle ownership, and road density (using data from IRF, 2009, and World Bank, 2013). Using this relationship, and the explanatory variables, kilometers driven per car are then derived for countries for which direct data are missing.

The natural logarithm of kilometers driven per car and the four explanatory indicators (of the 69 countries that have complete data) was taken and the second and third power of the log explanatory variables were included to add more flexibility. The regression results are shown in Annex Table 5.3.2, though again, because the equation is used for prediction, the interpretation of the estimated coefficients is not especially of concern.

ANNEX 5.4. ACCOUNTING FOR DELAYS TO ALL VEHICLE OCCUPANTS

This annex presents illustrative calculations to show how congestion cost estimates change when the mix of cars, buses, trucks, and two-wheel motorized vehicles is taken into account (the formulas in Annex 5.1 assume cars are the only vehicles).

Following from equation (5.1), the total costs of travel delays to all road users, when accounting for different vehicle types, is given by:

$$TC = \left(T - T^f\right) \times \sum_i V_i \times o_i \times VOT_i. \qquad (5.7)$$

ANNEX TABLE 5.3.2

Regression Results for Kilometers Driven per Car

Variables	Log km driven per car
log GDP per capita	−5.545*
	−3.058
log cars per capita	2.596**
	−1.127
log road length per car	3.359
	−2.091
log road density	0.113
	−0.16
log GDP per capita^2	−1.373**
	−0.66
log cars per capita^2	1.470***
	−0.487
log road length per car^2	0.992
	−0.738
log road density^2	−0.145
	−0.093
log GDP per capita^3	−0.093**
	−0.044
log cars per capita^3	0.197***
	−0.063
log road length per car^3	0.104
	−0.083
log road density^3	−0.016
	−0.033
Constant	−4.988
	−4.568
Observations	69
R-squared	0.642

Source: See main text.
Note: *, **, and *** indicate significance at the 10 percent, 5 percent, and 1 percent levels, respectively.

Subscript i is used to denote a particular type of vehicle: i = car, bus, truck, or two-wheeler. For simplicity, congestion is assumed to increase delay for all vehicles by the same absolute amount.

Differentiating equation (5.7) with respect to V_{car}, and using the definition of AC from equation (5.2), gives the following:

$$\frac{dTC}{dV_{car}} = AC + \frac{dT}{dV_{car}} \sum_i V_i \times o_i \times VOT_i. \tag{5.8}$$

Comparing equations (5.3) and (5.8), the ratio of the cost imposed on other vehicle occupants when there is a mix of vehicles as opposed to just cars is given by

$$\frac{\sum_i \left(\frac{V_i}{V}\right) \times o_i \times VOT_i}{o_{car} \times VOT_{car}}, \tag{5.9}$$

in which V_i/V is the share of vehicle i in total kilometers driven by all vehicles.

For the calculations in the remainder of this annex, the occupancy of trucks and two-wheelers is assumed to be one, and (based approximately on Parry and Small, 2009, for cities in the United Kingdom and the United States) that for buses is assumed to be 20. The VOT for two-wheelers and bus riders is assumed to be the same as for car occupants. For freight travel by trucks, the VOT should include the employer wage (the market wage plus employer payroll taxes) to reflect the per hour costs of labor time lost from congestion. Given that the VOT for car travel is 60 percent of the market wage, this implies $VOT_{truck}/VOT_{car} = 1.67$.

The last column of Annex Table 5.4.1 shows, based on equation (5.9), congestion costs with different scenarios for the vehicle fleet mix relative to congestion costs when cars are the only vehicles on the road.

If the only other vehicles are trucks and two-wheelers, there is relatively little difference in the results: in Annex Table 5.4.1, congestion costs are increased 1 percent when trucks account for 20 percent of the fleet and are reduced 7 percent when two-wheelers account for 20 percent of the fleet (with, in both cases, cars accounting for the remaining 80 percent). However, when buses account for 10 percent of the vehicle fleet congestion costs more than double, and when they account for 20 percent, costs more than triple. A car driver imposes considerably higher costs on others when the average number of vehicle occupants is higher, resulting from a significant share of high-occupancy buses on the roads. Congestion costs and motor fuel taxes may therefore be substantially understated in countries in which buses account for a significant share of vehicle traffic in urban centers.

ANNEX TABLE 5.4.1

Ratio of Congestion Cost with Multiple Vehicles Relative to Costs when Cars are the Only Vehicle

Share of vehicle-km by mode				Ratio of congestion cost with multiple vehicles to cost with cars only
Car	Bus	Truck	Two-wheeler	
1	0	0	0	1
0.8	0	0.2	0	1.01
0.8	0	0	0.2	0.93
0.8	0.2	0	0	3.30
0.5	0.1	0.1	0.3	2.04
0.9	0.1	0	0	2.15

Source: See text of Annex 5.4.

ANNEX 5.5. ASSESSMENT OF CONGESTION COSTS FROM COUNTRY-LEVEL DATA: THE UNITED STATES AND THE UNITED KINGDOM

This annex explains the supplementary estimation of delays at the country level for the United States and the United Kingdom, mentioned in the main text. Estimated delays are for 2008 (as a close approximation to 2010) and costs are expressed in 2010 U.S. dollars.

The United States

For the United States, the Texas Transportation Institute (TTI) compiles high-quality data on travel delays for 449 urban centers categorized by population size into very large, large, medium, and small cities (Schrank, Lomax, and Eisele, 2011).

For the 101 largest cities, speed data are collected remotely by a private company for different times of the day for each link within the urban road network. For the other 348 smaller urban centers (which account for 15 percent of nationwide travel delays), speed is derived from estimated speed/traffic volume relationships. Schrank, Lomax, and Eisele (2011) use traffic volume data from the Highway Performance Monitoring System, an inventory maintained by the Federal Highway Administration for all roadway segments in the United States.

The TTI report for 2008 was used to derive the nationwide congestion delay on others. For each urban region in the TTI sample, total annual hours of delay to passengers in cars is divided by total annual vehicle-kilometers driven by cars to give the average hourly delay per car-kilometer. Delays at the regional level are weighted by the share of car-kilometers in nationwide kilometers and then aggregated to obtain a nationwide average measure of delay.

The United Kingdom

For the United Kingdom, travel data for 2008 were obtained from the U.K. Department for Transport (DFT), which compiles official statistics on the British transport system. Because DFT does not provide annual hours of travel delays at the city level, travel delays were generated by comparing average vehicle speed during peak times with the free flow speed, both of which can be obtained from the DFT statistics.[28]

For each of the U.K. localities, the average travel time per kilometer was calculated along with the free flow average travel time per kilometer, using the average travel speed during morning peak (7 am to 10 am) and the free flow speed.

---

[28]The data used are from http://www.dft.gov.uk/statistics/tables, specifically data sets CGN0201, SPE0104, TRA8901, and TRA0307.

Annual car-kilometers within each locality were then multiplied by the share of car-kilometers occurring during the morning peak period. The total annual hours of delay was then obtained by multiplying the extra travel time per kilometer during morning peak time by 2 to account for the evening peak (4 pm to 7 pm) which is assumed to experience the same traffic congestion.[29]

Next, the total annual hours of delay were divided by total annual vehicle-kilometers driven by cars for each locality to derive the average hourly delay per vehicle-kilometer, which was then converted into passenger delays assuming an average vehicle occupancy of 1.6. Average delay at the nationwide level is a weighted average of that at the locality level, with weights equal to the shares in nationwide car-kilometers.

Delays to others per car-kilometer are about twice as high for the United Kingdom as for the United States, which seems roughly plausible, given that a much greater share of nationwide driving occurs under congested conditions in the United Kingdom.

ANNEX 5.6. ESTIMATING THE RATIO OF OTHER ACCIDENT COSTS TO FATALITY COSTS: COUNTRY CASE STUDIES

As mentioned in the main text, external accident costs for different countries are obtained by scaling up estimates of external fatality costs by the ratio of other costs to external fatality costs. This ratio—which is based on several country case studies—is attained as follows:

First, using data compiled by Herrnstadt, Parry, and Siikamäki (2013) for Finland, Sweden, the United Kingdom, and the United States and by Parry and Strand (2012) for Chile, comprehensive estimates of external accident costs were made for these five countries for 2010 or the latest possible. In these calculations, external fatalities were monetized using mortality values discussed in Chapter 4. Other costs were valued using a combination of local data on the average (personal, medical, and property damage) costs associated with accidents of different severity, and in some cases, extrapolations of these costs from U.S. data.[30] Approximately 85 percent of medical costs are assumed to be external (borne by third parties) for all fatal and nonfatal and internal and external injuries, and 50 percent of property damage costs (for all accidents) are external.

From these studies, five point estimates for the ratio of other external costs (medical, property damage, and nonfatal injury costs) to external fatality costs were

---

[29]According to DFT traffic distribution data (TRA0307), the shares of morning and evening peak kilometers in total kilometers driven are 0.21 and 0.22, respectively, which are very close.
[30]Detailed documentation of data sources and estimation procedures are provided in the above references. The breakdown of fatalities by driver, passenger of drivers, other vehicle occupant, pedestrian, and cyclist is available from the data sources, and this breakdown is assumed to be the same for nonfatal injuries. Herrnstadt, Parry, and Siikamäki (2013) focus only on alcohol-related accidents; their data were modified to include data for all traffic accidents.

obtained. This ratio tends to decline as the relative importance of pedestrian and other external deaths in total road deaths rises (the numerator in the ratio falls and the denominator rises). This ratio is 2.9 in the United States (where 23 percent of deaths are external) and only 0.16 in Chile (where 54 percent of deaths are external). A power function that best fits these five data points relating this cost ratio to the share of external fatalities in total fatalities was estimated.[31] This relationship was then used to extrapolate the other-external-cost-to-fatality-external-cost ratio for other countries depending on their share of external fatalities in total fatalities.

ANNEX 5.7. MISCELLANEOUS DATA AND PROCEDURES FOR CALCULATING CORRECTIVE TAXES

The remaining data and assumptions needed to implement the corrective fuel tax formula as set forth in Chapter 3 are discussed in this annex. These issues deal with the use of diesel fuel by both cars and trucks; the breakdown of fuel price responses; and the conversion of road damage, accident, local pollution, and congestion costs into corresponding components of corrective fuel taxes. To make this conversion, fuel efficiency is needed to convert any road damage, accident, local pollution, or congestion costs expressed per vehicle-kilometer driven into a cost per liter of fuel. However, given the difficulty of accurately measuring fuel efficiency for most countries (see below), these costs are directly expressed per liter insofar as possible, to avoid the need for this data.

Diesel use by different vehicle types: External costs for cars are used to calculate corrective taxes on gasoline. However, diesel fuel is used by both cars and trucks and, given the practical difficulty of differentiating the diesel tax according to vehicle use, a weighted average of external costs for cars and trucks should be used in the corrective diesel tax formula, based on their respective shares in diesel fuel consumption. The breakout of diesel fuel use by cars versus trucks is available for a limited number of countries and for other countries was taken from regional average figures.[32]

Breakdown of fuel price responses: An important piece of data is the fraction of the fuel demand response that comes from reduced driving (as opposed to the remaining fraction that comes from fuel efficiency improvements). For cars, this fraction is assumed to be 0.5 for all countries.[33] For diesel fuel used by trucks

---

[31]Specifically, the cost ratio is predicted by the equation $0.049x^{-2.56}$, in which x is the share of external fatalities in total fatalities.

[32]The data source is ICCT (2010). For example, cars account for about 11 percent of road diesel at the global level, and 32 percent in the European Union.

[33]See Small and Van Dender (2006) and the review of other studies in Parry and Small (2005). In practice this fraction will vary across countries; for example, it might be higher in countries with readily available alternatives to car use (which increases the responsiveness of driving to fuel prices) and in countries with binding fuel efficiency regulations (which reduces the responsiveness of fuel efficiency to fuel prices). However, no international data on which country-specific assumptions can be based are available.

(where the high power requirements necessary to move freight limit opportunities for improving fuel efficiency through, for example, reducing vehicle size and weight) this fraction is assumed to be 0.6 (Parry, 2008).

Road damage: The estimation procedures outlined in the main text yield total external costs for road damage. These costs are divided by total diesel fuel consumption for trucks to obtain costs per liter, which are then multiplied by 0.6 to account for the portion of the fuel response that comes from reduced kilometers driven.

Accidents: The estimation procedures also yield total external costs for traffic accidents. However, expressing them per liter of fuel is more involved because external costs per vehicle-kilometer are assumed to be the same for cars and trucks, implying external costs per liter of fuel will be larger for cars than for trucks given that cars travel farther on a liter of fuel, and larger for diesel fuel cars than gasoline cars (because diesel cars are more fuel efficient). Truck fuel efficiency is assumed to be one-third that for gasoline cars (Parry, 2008), and in turn, diesel cars are assumed to be 20 percent more fuel efficient than their gasoline counterparts.

External accident costs per liter of gasoline can then be obtained by dividing total accident costs for all vehicles by a weighted sum of fuel use by gasoline cars, diesel cars, and trucks (fuel use data are discussed in the Annex 6.1); the weights are fuel efficiency of other vehicles relative to that for gasoline cars (1.2 and 0.33, respectively, for diesel cars and trucks). In turn, costs per liter for diesel cars and trucks are the external costs per liter for gasoline cars multiplied by the same weights. In applying the costs to the corrective fuel tax formula, they are again multiplied by the portion (0.5 or 0.6) of the fuel price response that comes from reduced driving as opposed to fuel efficiency improvements.

Local pollution: Local pollution damage is estimated on a per liter basis. The scaling factor, however, depends on how emissions are regulated. In countries such as the United States, where emissions are regulated on a per kilometer (or per mile) basis and approximately maintained throughout the vehicle's life, roughly speaking emissions vary only with kilometers driven, not fuel efficiency, and therefore need to be multiplied by the driving fraction of the fuel price response.[34] In countries with less effective regulation, emission might be appropriately proportional to fuel use. The calculations in this analysis apply a scaling factor of 0.5 (gasoline vehicles) or 0.6 (diesel vehicles) for Australia, Canada, China, European countries, New Zealand, Singapore, and the United States, and 1.0 (no adjustment) for all other countries. More refined assumptions would not have that much effect on the corrective fuel tax estimates given the relatively large size of congestion and accident costs (see Chapter 6).

---

[34]In some cases, emissions standards are defined with respect to engine capacity (e.g., in European Union countries as well as in other countries adopting European Union standards). In this analysis, some fuel efficiency improvements, such as reducing vehicle weight, will affect emissions but others, such as more efficient engines, will not.

Congestion: Congestion costs are estimated on a per kilometer basis and therefore need to be multiplied by fuel efficiency (see below) to express them in per liter terms (after scaling by the driving fraction of the fuel price response).

One complication is that driving on congested roads (mostly by people commuting to work) is generally less sensitive to fuel prices than driving on uncongested roads. This fact reduces the congestion benefits from higher fuel taxes. Based on evidence of the relative price responsiveness of driving under congested and noncongested conditions, Parry and Small (2005) recommend scaling back congestion costs by a third in computing corrective fuel taxes; the same procedure is followed here.

Fuel efficiency: Fuel efficiency (of vehicles in use on the road) could be obtained by dividing data on vehicle-kilometers driven by fuel use. However, because the reliability of the vehicle-kilometer data varies across countries (being generally less accurate for developing countries), fuel efficiency is based instead on a plausible assumption for different regions, and applied to all countries in the region. For example, based on estimates in Parry and Small (2005) for the United States and the United Kingdom, fuel efficiency for gasoline vehicles is assumed to be 10.5 kilometers/liter (25 miles/gallon) in North America and 14.5 kilometers/liter (35 miles/gallon) for higher-income European countries and Japan.[35] Fuel efficiency for diesel cars and trucks is then derived using the above ratios (1.2 and 0.33, respectively). Other assumptions would moderately affect the contribution of congestion costs to corrective taxes.

Finally, to simplify the computation of corrective fuel taxes, it is assumed that fuel efficiency in each country remains fixed at its current level, rather than increasing in response to higher fuel prices. This assumption leads again to some understatement of the corrective fuel tax because, per liter of fuel reduction, the reduction in vehicle-kilometers driven (and hence congestion and accidents) is greater for a more fuel-efficient vehicle; see equation (3.1).[36]

REFERENCES

Abrantes, P.A.L., and M.R. Wardman, 2011, "Meta-Analysis of UK Values of Travel Time: An Update," *Transportation Research Part A: Policy and Practice*, Vol. 45, No. 1, pp. 1–17.

Becker, Gary S., 1965, "A Theory of the Allocation of Time," *The Economic Journal*, Vol. 75, pp. 493–511.

Bureau of Transportation Statistics (BTS), 2012, *National Transportation Statistics* (Washington: Bureau of Transportation Statistics, US Department of Transportation).

Commissariat Général du Plan, 2001, *Transports: Choix des Investissements et Coût des Nuisances [Transportation: Choice of Investments and the Cost of Nuisances]* (Paris).

Duranton, Gilles, and Matthew A. Turner, 2011, "The Fundamental Law of Road Congestion: Evidence from US Cities," *American Economic Review*, Vol. 101, pp. 2616–52.

---

[35]These figures make some adjustment for recent increases in fuel efficiency. Other assumptions are Central and South America and Eurasia, 10.5 kilometers/liter; lower-income Europe and Asia, 12.5 kilometers/liter; and Middle East, 8.5 kilometers/liter.

[36]The understatement is not huge, however. For example, based on assumptions here and in Annex 6.1, even a 50 percent increase in gasoline prices would increase fuel efficiency by 12.5 percent (see also Small and Van Dender, 2006).

Edlin, Aaron S., and Pinar Karaca-Mandic, 2006, "The Accident Externality from Driving," *Journal of Political Economy*, Vol. 114, pp. 931–55.

Eliasson, Jonas, 2006, "Forecasting Travel Time Variability," Proceedings of the European Transport Conference (Henley-in-Arden, U.K.: Association for European Transport).

Fosgerau, M., K. Hjorth, C. Brems, and D. Fukuda, 2008, *Travel Time Variability: Definition and Valuation* (Lyngby, Denmark: DTU Transport).

Fridstrøm, Lasse, Jan Ifver, Siv Ingebrigtsen, Risto Kulmala, and Lars Krogsgård Thomsen, 1995, "Measuring the Contribution of Randomness, Exposure, Weather, and Daylight to the Variation in Road Accident Counts," *Accident Analysis and Prevention*, Vol. 27, pp. 1–20.

Greenwood, Ian D., and Christopher R. Bennett, 1996, "The Effects of Traffic Congestion on Fuel Consumption," *Road and Transport Research*, Vol. 5, pp. 18–31.

Herrnstadt, Evan, Ian W.H. Parry, and Juha Siikamäki, 2013, "Do Alcohol Taxes in Europe and the US Rightly Correct for Externalities?" *International Tax and Public Finance*, published online September 25.

International Council on Clean Transportation (ICCT), 2010, Global Transportation Roadmap Model (Washington: International Council on Clean Transportation).

International Labor Organization (ILO), 2012, Global Wage Database 2012 (Geneva: International Labor Organization). www.ilo.org/travail/areasofwork/wages-and-income/WCMS_142568/lang—en/index.htm.

International Road Federation (IRF), 2009, *World Road Statistics 2009* (Geneva: International Road Federation).

———, 2012, *World Road Statistics 2012* (Geneva: International Road Federation).

Kopits, Elizabeth, and Maureen Cropper, 2008, "Why Have Traffic Fatalities Declined in Industrialized Countries? Implications for Pedestrians and Vehicle Occupants," *Journal of Transport Economics and Policy*, Vol. 42, pp. 129–54.

Lindberg, Gunnar, 2001, "Traffic Insurance and Accident Externality Charges," *Journal of Transport Economics and Policy*, Vol. 35, pp. 399–416.

Lindsey, Robin, 2010, "Dedicated Lanes, Tolls and Its Technology," in *The Future of Interurban Passenger Transport* (Paris: Organization for Economic Cooperation and Development).

Mackie, P.J., M. Wardman, A.S. Fowkes, G. Whelan, J. Nellthorp, and J. Bates, 2003, "Values of Travel Time Savings in the UK: Summary Report" (Leeds, U.K.: Institute for Transport Studies, University of Leeds).

Newbery, David M., 1987, "Road User Charges in Britain," *Economic Journal*, Vol. 98, No. 390 (Supplement), pp. 161–76.

———, 1990, "Pricing and Congestion: Economic Principles Relevant to Pricing Roads," *Oxford Review of Economic Policy*, Vol. 6, pp. 22–38.

Parry, Ian W.H., 2004, "Comparing Alternative Policies to Reduce Traffic Accidents," *Journal of Urban Economics*, Vol. 56, pp. 346–68.

———, 2008, "How Should Heavy-Duty Trucks be Taxed?" *Journal of Urban Economics*, Vol. 63, pp. 651–68.

———, and Kenneth A. Small, 2005, "Does Britain or the United States Have the Right Gasoline Tax?" *American Economic Review*, Vol. 95, No. 4, pp. 1276–89.

———, 2009, "Should Urban Transit Subsidies Be Reduced?" *American Economic Review*, Vol. 99, pp. 700–24.

Parry, Ian W.H., and Jon Strand, 2012, "International Fuel Tax Assessment: An Application to Chile," *Environment and Development Economics*, Vol. 17, pp. 127–44.

Paterson, William D.O., 1987, *Road Deterioration and Maintenance Effects—Models for Planning and Management*, World Bank Highway Design and Maintenance Standards Series (Baltimore, Maryland: Johns Hopkins University Press).

Peer, S., C.C. Koopmans, and E.T. Verhoef, 2012, "Prediction of Travel Time Variability for Cost-Benefit Analysis," *Transportation Research Part A: Policy and Practice*, Vol. 46, pp. 79–90.

Schrank, David, Tim Lomax, and Bill Eisele, 2011, *2011 Urban Mobility Report* (College Station, Texas: Texas Transportation Institute, Texas A&M University).

Small, Kenneth A., 1992, *Urban Transportation Economics*. In Fundamentals of Pure and Applied Economics Series, vol. 51 (Chur, Switzerland: Harwood Academic Press).
Small, Kenneth A., and Jose A. Gómez-Ibáñez, 1998, "Road Pricing for Congestion Management: The Transition from Theory to Policy," in *Road Pricing, Traffic Congestion and the Environment: Issue of Efficiency and Social Feasibility*, edited by K.J. Button and E.T. Verhoef (Cheltenham, U.K.: Edward Elgar) pp. 213–46.
Small, Kenneth A., and Kurt Van Dender, 2006, "Fuel Efficiency and Motor Vehicle Travel: The Declining Rebound Effect," *Energy Journal*, Vol. 28, No. 1, pp. 25–52.
Small, Kenneth A., and Erik Verhoef, 2007, *The Economics of Urban Transportation* (New York: Routledge).
Small, Kenneth A., Clifford Winston, and Carol A. Evans, 1989, *Road Work* (Washington: Brookings Institution).
Transport Canada, 1994, *Guide to Benefit-Cost Analysis in Transport Canada* (Ottawa, Ontario: Transport Canada).
———, 2006, *Costs of Non-Recurrent Congestion in Canada*. Final Report. TP 14664E (Ottawa, Ontario: Transport Canada).
Transportation Research Board, 2010, *Highway Capacity Manual 2010* (Washington: National Academies). http://pereview.net/wp-content/uploads/pdf/hcm-extracts.pdf.
Traynor, Thomas L., 1994, "The Effects of Varying Safety Conditions on the External Costs of Driving," *Eastern Economic Journal*, Vol. 20, pp. 45–60.
United States Department of Transportation, 1997, *The Value of Travel Time: Departmental Guidance for Conducting Economic Valuations* (Washington: US Department of Transportation).
United States Federal Highway Administration (US FHWA), 1997, *1997 Federal Highway Cost Allocation Study* (Washington: Federal Highway Administration, US Department of Transportation).
———, 2005, *Crash Cost Estimates by Maximum Police-Reported Injury Severity Within Selected Crash Geometries*, FHWA-HRT-05-051 (Washington: Federal Highway Administration, US Department of Transportation).
Wardman, Mark, 1998, "The Value of Travel Time: A Review of British Evidence," *Journal of Transport Economics and Policy*, Vol. 32, pp. 285–316.
———, 2001, "A Review of British Evidence on Time and Service Quality Valuations," *Transportation Research E*, Vol. 37, pp. 107–28.
Waters, William G. II, 1996, "Values of Time Savings in Road Transport Project Evaluation," in *World Transport Research: Proceedings of 7th World Conference on Transport Research*, vol. 3., edited by David Hensher, J. King, and T. Oum (Oxford, U.K.: Pergamon) pp. 213–23.
World Bank, 2013, World Development Indicators Database (Washington: World Bank). http://data.worldbank.org/indicator.
World Health Organization (WHO), 2013, Global Health Observatory Data Repository. (Geneva: World Health Organization). http://apps.who.int/gho/data/node.main.A997?lang = en.

CHAPTER 6

The Right Energy Taxes and Their Impacts

This chapter summarizes the corrective tax estimates for coal, natural gas, and motor fuels based on the assumptions discussed in previous chapters, both for selected countries and, using ranges of values in heat maps, for all countries, and then discusses the fiscal, health, and environmental impacts of tax reform. Various tables in Annex 6.2 provide full details of this information, country by country, including estimates of current fuel taxes or subsidies.

CORRECTIVE TAX ESTIMATES

Coal

Figure 6.1 illustrates the corrective taxes (taxes that reflect environmental damage) on coal use by power plants from a representative sample of countries with different income levels, geographical locations, and energy mixes. The figure shows the tax on coal necessary to correct for carbon emissions (based on the illustrative damage value of $35/ton of carbon dioxide [CO_2]) and the additional taxation needed to reflect local pollution damage, based on the average emission rates at existing plants. Current taxes as shown in the figure, obtained from Clements and others (2013),[1] are approximately zero. To put the corrective tax estimates in perspective, the world average coal price in 2010 was about $5 per gigajoule (GJ).[2]

A number of noteworthy points can be made from Figure 6.1. The carbon component of the corrective tax is substantial, equivalent to about $3.3/GJ, or about 66 percent of the average world coal price in 2010. The corrective tax for carbon varies little across countries because there is little variation in carbon emissions/GJ of coal, and the illustrated CO_2 damage value is applied to all countries.

More striking, however, is that the local pollution charge is often larger than the carbon charge, though it varies considerably across countries. The local pollution charge exceeds the carbon charge for 10 countries shown in the figure and is more than double the carbon charge in 6 of those countries. For example, in

---

[1] These taxes are calculated on a consistent basis across countries, based primarily on a comparison of domestic fuel prices with international prices.
[2] This and other global average fuel prices mentioned in this chapter are also from Clements and others (2013).

Figure 6.1 Corrective Coal Tax Estimates, Selected Countries, 2010

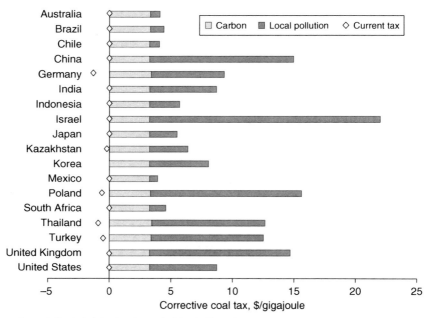

Source: Authors' calculations based on methodology in Chapter 4.
Note: The dark gray bar is the corrective charge for local pollution based on emission rates averaged across existing plants (some of which have control technologies and some of which do not). A current tax data point to the left of the y-axis indicates a subsidy. Data on Korea's coal tax were not available from the sources used in this book.

the United States local pollution damage contributes $5.5/GJ to the corrective tax, or 62 percent of the total corrective coal tax of $8.7/GJ. For China, the corrective charge for local pollution is $11.7/GJ; in fact, premature deaths per ton of coal burned in China are about nine times those for the United States (due to a combination of higher average emission rates and higher population exposure), though a partially offsetting factor is that, because of lower per capita income, the value of mortality risk is assumed to be lower in China (see Figure 4.2).

Corrective taxes for local pollution are not always large. For example, in Australia the corrective charge is $0.8/GJ, which in part reflects relatively low population exposure to the pollution (much of which disperses over the ocean).

Figure 6.2 underscores the potentially strong incentives for plants to adopt emission control technologies when faced with high air pollution charges on coal with appropriate crediting for the use of control technologies. It shows the corrective air pollution charges for existing plants with emissions control technologies and those for plants with no controls. (The average emissions levels underlying the corrective charges in Figure 6.1 are based on intermediate emission rates.) Taxes are reduced by 75 percent or more by adopting control technologies in all cases. However, even if the emissions control technologies currently used by some plants in a country were applied to all plants, corrective taxes for air pollution

Figure 6.2 Corrective Taxes for Air Pollution at Coal Plants with and without Control Technologies, Selected Countries, 2010

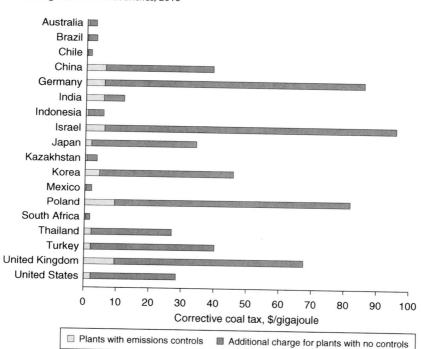

Source: Authors' calculations based on methodology in Chapter 4.
Note: The sum of the light and dark gray bars is the appropriate coal tax for local air emissions for plants with no emissions controls. The light gray bar is the tax that would be paid for a representative plant with control technologies receiving an appropriate credit for the emissions mitigation.

could still be significant—the taxes would be greater than the carbon charge for seven of the countries shown in Figure 6.1.[3]

Figure 6.3 shows the breakdown of air pollution damage from coal plants with no controls by type of emissions. For most countries, SO_2 is the most damaging pollutant (its share in total pollution damage varies across countries from 27 percent to 71 percent), followed by primary fine particulate emissions ($PM_{2.5}$), though in some countries (Australia, Brazil, India, Japan, and Korea) primary particulates from uncontrolled plants would cause the most damage (their share in total pollution damage varies from 16 percent to 66 percent). Nitrogen oxide emissions are responsible for a relatively minor share of damage (2–16 percent) because their

---

[3] Governments frequently use regulatory approaches to promote the use of control technologies but, besides raising revenue, fiscal instruments likely provide more robust incentives. As discussed in Box 3.1, fiscal instruments provide greater incentives to shift to coal with lower pollution content or to other generation fuels, and to reduce electricity demand.

Figure 6.3 Breakdown of Air Pollution Damages from Coal by Emissions Type, Selected Countries, 2010

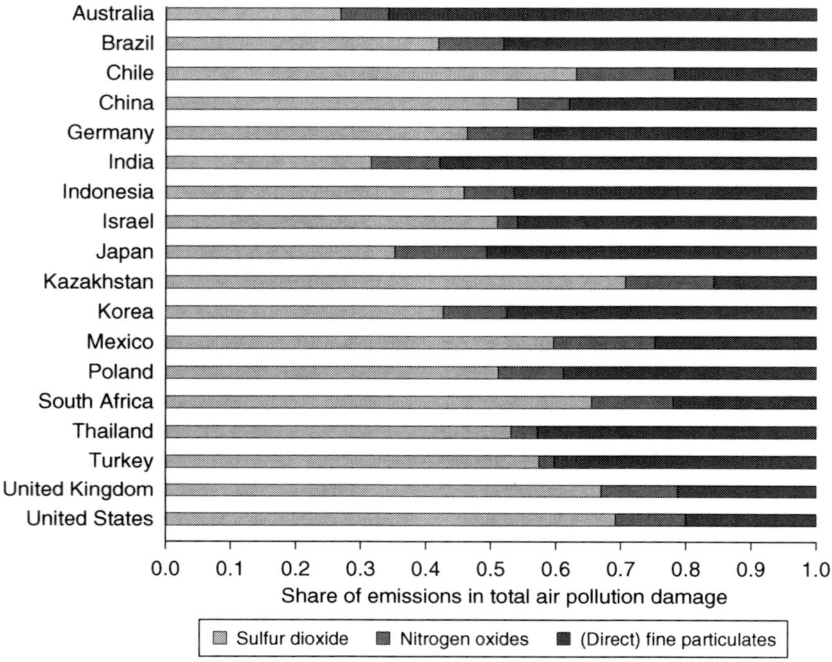

[Horizontal stacked bar chart showing share of emissions in total air pollution damage (0.0 to 1.0) for: Australia, Brazil, Chile, China, Germany, India, Indonesia, Israel, Japan, Kazakhstan, Korea, Mexico, Poland, South Africa, Thailand, Turkey, United Kingdom, United States. Legend: Sulfur dioxide, Nitrogen oxides, (Direct) fine particulates.]

Source: Authors' calculations based on methodology in Chapter 4.

emission rates are smaller than for SO_2, and they are less prone to reacting in the atmosphere to form the fine particulates that give rise to major health risks.

Figure 6.4 shows a heat map of corrective coal tax estimates, based on average or current emission rates, for all countries that use coal. The relative cross-country pattern of corrective taxes looks broadly similar to that for sulfur damage presented in Chapter 4 (though Figure 6.3 also reflects other local pollutants, carbon damage, and the emissions intensity of locally used coal). Corrective taxes tend to be high in Europe (where population exposure and per capita income are relatively high) and lowest in the limited number of African countries that use coal and for which data are available, with countries in North and South America, Asia, and Oceania generally in between.

Figure 6.5 reproduces the corrective coal tax estimates for selected countries from Figure 6.1, but uses the same mortality value (that for the average Organization for Economic Cooperation and Development [OECD] country) in all cases, given the controversy about applying different values to different countries. Not surprisingly, the main effect is to substantially scale up the corrective tax estimates for countries with per capita incomes well below the OECD average. For example, China's corrective tax for air pollution increases from $11.7/GJ to $38.7/GJ.

The Right Energy Taxes and Their Impacts | 135

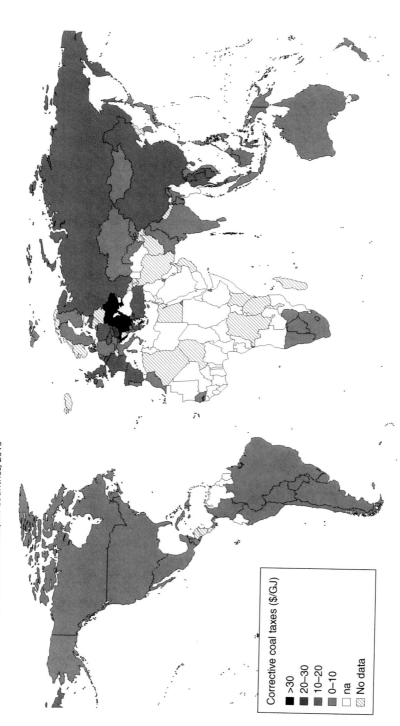

Figure 6.4 Corrective Coal Tax Estimates, All Countries, 2010

Corrective coal taxes ($/GJ)
- ■ >30
- ▓ 20–30
- ▒ 10–20
- ░ 0–10
- □ na
- ▨ No data

Source: Authors' calculations based on methodology in Chapter 4.

Figure 6.5 Corrective Coal Tax Estimates with Uniform Mortality Values, Selected Countries, 2010

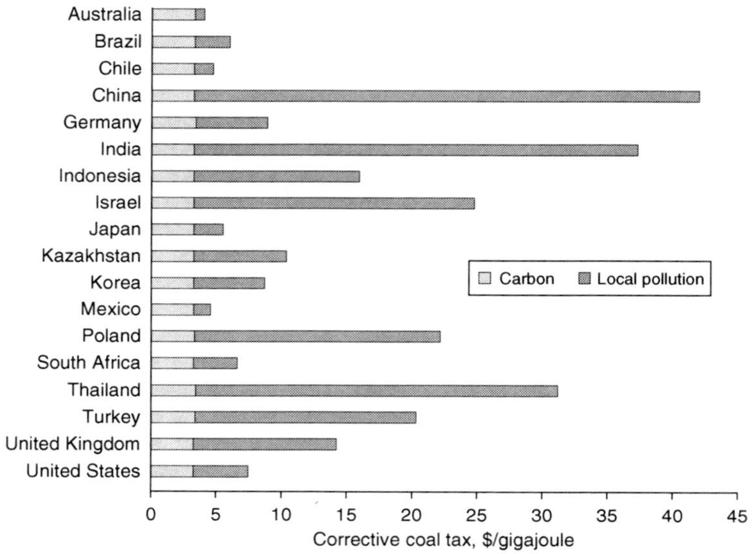

[Bar chart showing corrective coal tax in $/gigajoule, split between Carbon and Local pollution, for the following countries:
- Australia
- Brazil
- Chile
- China (~42)
- Germany
- India (~37)
- Indonesia
- Israel
- Japan
- Kazakhstan
- Korea
- Mexico
- Poland
- South Africa
- Thailand (~31)
- Turkey
- United Kingdom
- United States
X-axis: 0 to 45, Corrective coal tax, $/gigajoule]

Source: Authors' calculations based on methodology in Chapter 4.
Note: This figure adjusts the corrective tax estimates for air pollution from Figure 6.1 by setting mortality values for all countries equal to the average value for Organization for Economic Cooperation and Development countries.

Dramatic differences in corrective taxes across countries still remain, however, reflecting large differences in population exposure and emission rates.

Natural Gas

Figure 6.6 shows taxes for natural gas to correct for carbon and air pollution damage at the average power plant. Current taxes are about zero in many cases, though natural gas is subsidized in some countries, especially in Egypt ($1.4/GJ) and India ($1.0/GJ). For perspective, the world average price for pipeline natural gas in 2010 was about $5/GJ.

Figures 6.6 and 6.1 illustrate the significant differences between coal and natural gas. First, the carbon charge is much lower for natural gas, about $1.9/GJ or about 55 percent of that for coal. This lower charge reflects the lower carbon emission rate per GJ of energy for natural gas.

Second, local pollution damage is also much lower. For all but one country (Korea) the corrective tax for local air pollution is less than the carbon charge, and often very much smaller: in 7 of the 20 countries shown the corrective tax for local pollution is less than 10 percent of that for carbon. Gas combustion generates minimal amounts of SO_2 and $PM_{2.5}$—the two largest sources of air pollution damage for coal. Moreover, the nitrogen oxide emission rates per GJ for natural gas are less than half of those for coal.

Although less dramatic than for coal, there is still significant undercharging for natural gas, with currently estimated taxes for the countries shown in Figure 6.6

Figure 6.6 Corrective Natural Gas Tax Estimates for Power Plants, Selected Countries, 2010

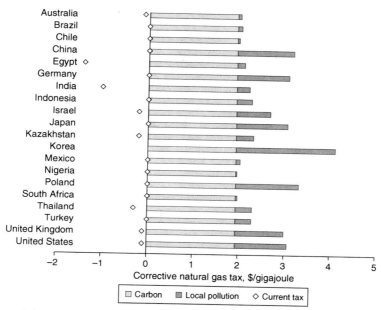

Source: Authors' calculations based on methodology in Chapter 4.
Note: Data on Korea's natural gas tax were not available from the sources used in this book. A current tax data point to the left of the y-axis indicates a subsidy.

either about zero or negative, compared with corrective charges of about 40 percent or more of the world price.

Figure 6.7 shows corrective tax estimates for natural gas used at power plants across all countries. The differences across countries are far less pronounced than for coal given that local pollution damage is much smaller for natural gas relative to coal.

The corrective taxes for natural gas combusted at ground level, such as for home heating, are similar to those for power plant use of natural gas and, again, the charge for local air pollution is less important (see Annex 6.1). Given the dominance of the carbon charge in each case, there does not appear to be a strong case, on pollution grounds, for differentiating natural gas taxes by end user.

Motor Fuels

Figure 6.8 shows estimates of corrective gasoline taxes for selected countries expressed in 2010 US$/liter,[4] and the contribution of carbon, local pollution, traffic accidents, and congestion to the corrective tax. The corrective tax refers to the excise tax before application of any value-added or sales taxes. Current estimated excise taxes vary considerably, from subsidies of US$0.30/liter in Egypt to taxes of US$0.60/liter or more in Brazil, Germany, Israel, Japan, Korea, Poland,

---

[4] To convert to dollars per gallon, multiply by 3.8.

Figure 6.7 Corrective Natural Gas Tax Estimates for Power Plants, All Countries, 2010

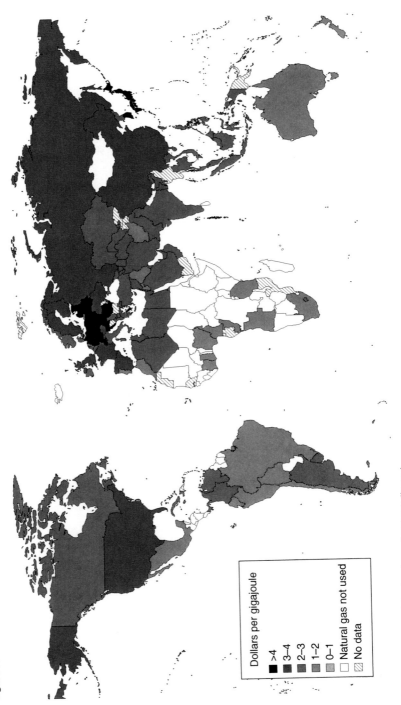

Source: Authors' calculations based on methodology in Chapter 4.

Figure 6.8 Corrective Gasoline Tax Estimates, Selected Countries, 2010

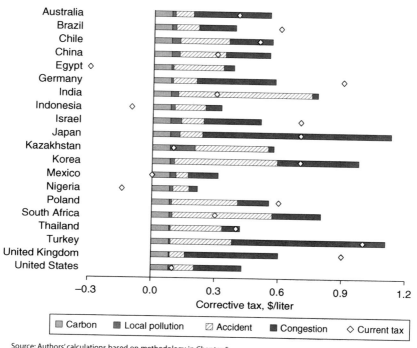

Source: Authors' calculations based on methodology in Chapter 5.
Note: To express taxes in dollars per gallon, multiply by 3.8.

Turkey, and the United Kingdom. For perspective, the world pretax price of gasoline averaged about US$0.80/liter in 2010, equivalent to $23/GJ.

The carbon component of the corrective gasoline tax, based on the damage assumption of $35/ton, is US$0.08/liter across all countries, or $2.4/GJ, a little higher than for natural gas, although for gasoline the carbon charge is a smaller portion (about 10 percent) of the world price.

As with natural gas the local pollution component is usually smaller than the carbon charge, and mainly for the same reason: gasoline produces only very small amounts of the most damaging pollutants—SO_2 and $PM_{2.5}$. Carbon and local pollution together point to corrective gasoline taxes of, at most, US$0.20/liter for the illustrated countries.

However, much heavier taxation of gasoline is warranted by other factors, taxes higher than currently imposed in most cases, because of the combination of traffic congestion and traffic accidents. Traffic congestion tends to be the largest component of the corrective tax in developed countries, in part because of higher values from lost time, and traffic accidents are the largest component in developing countries (where, for example, pedestrians are more prone to injury risk). These additional costs raise corrective gasoline taxes in Figure 6.8 to between

140 Getting Energy Prices Right: From Principle to Practice

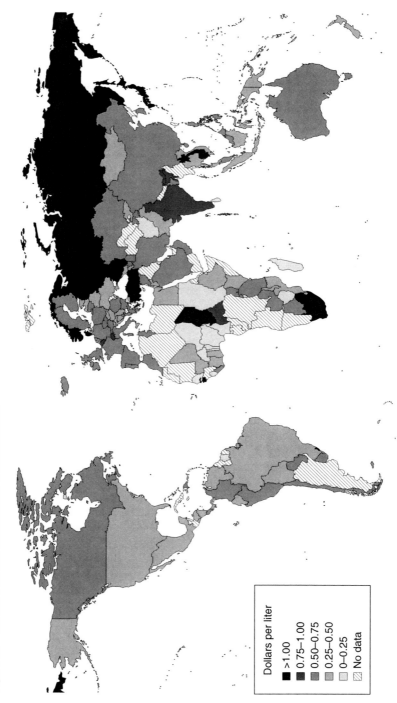

Figure 6.9 Corrective Gasoline Tax Estimates, All Countries, 2010

Source: Authors' calculations based on methodology in Chapter 5.

about US$0.40 and US$0.60/liter in Australia, Brazil, Chile, China, Egypt, Germany, Israel, Kazakhstan, Poland, Thailand, and the United States and to about US$0.80/liter (or 100 percent of the pretax world price) or more in India, Japan, Korea, South Africa, and Turkey. The corrective tax estimates exceed current taxes for 15 of the countries in Figure 6.7, and fall short of them for five countries.[5]

Figure 6.9 underscores the heavy taxation of gasoline warranted worldwide. For the majority of countries for which data are available, corrective taxes are at least US$0.40/liter (50 percent of the pretax world price) and frequently much higher. Broadly speaking, the gasoline tax rates currently applicable in most OECD countries, about US$0.40 to $1/liter, appear to be in the right ballpark for countries worldwide (at least until distance-based charging becomes widespread).

Turning to diesel fuel, the corrective excise tax estimates for selected countries in Figure 6.10, averaged for diesel use by cars and trucks, follow a pattern broadly similar to those for gasoline taxes in Figure 6.7. Most estimates are between about

Figure 6.10 Corrective Diesel Tax Estimates, Selected Countries, 2010

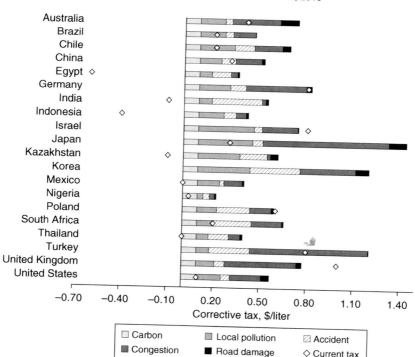

Source: Authors' calculations based on methodology in Chapter 5.
Note: To express taxes in dollars per gallon, multiply by 3.8. Data on Korea's diesel tax were not available from the sources used in this book.

---

[5]Even in these cases (e.g., Germany and the United Kingdom) reductions in gasoline taxes may not be efficient in practice because the corrective taxes may be understated for a variety of reasons (as discussed in Chapter 5).

US$0.40 and US$0.80/liter, though corrective taxes for Japan, Korea, and Turkey are much higher. For 15 countries, corrective diesel fuel taxes are somewhat higher than corrective gasoline taxes, suggesting that, if anything, diesel should be taxed more heavily than gasoline. Many governments, however, tax diesel at a lower rate than gasoline (10 of the countries shown in Figure 6.10 and 28 of 34 OECD countries shown in Figure 2.14). Corrective taxes for diesel exceed current taxes in all but two countries (Israel and the United Kingdom) in Figure 6.10.

The generally higher corrective taxes for diesel fuel compared with gasoline reflect its higher emission rates (both for carbon and, especially, local pollution); that most diesel is used by trucks, which add more to congestion per vehicle-kilometer than cars; and that trucks cause more road damage. However, road damage is relatively modest—less than the local pollution component in all cases. And a partially offsetting factor is that trucks have much lower fuel efficiency than cars, which means that a liter reduction in diesel fuel consumption results in a much smaller reduction in vehicle-kilometers driven (implying smaller congestion and accident benefits) compared with a liter reduction in car-kilometers driven.

In fact, if it were administratively feasible to do so, a case could be made for taxing diesel fuel used by cars at a higher rate than that used by trucks—the corrective tax for car diesel is higher than for truck diesel (these results are not shown in the figures), though the differences tend to be fairly modest.

IMPACTS

The findings in this chapter provide a useful marker for policymakers interested in heading toward the efficient system of fuel taxes needed to balance environmental and economic concerns. Obviously, however, the fiscal, health, and environmental impacts of tax reforms are of great interest in themselves—particularly for helping policymakers prioritize among different reform options. These impacts will vary considerably across countries depending, for example, on each country's prevailing fuel mix and fiscal and other policies currently affecting energy and transportation systems.[6]

Tax reform options are now compared based on "back-of-the-envelope" calculations described in Annex 6.1. This comparison involves estimating the change in fuel prices that would result from implementing corrective taxes (relative to current taxes, which are often zero and sometimes negative, and assuming full pass-through into consumer prices). The price changes are combined with an assumption, based loosely on the limited evidence available, that each 1 percent increase in a fuel price eventually reduces use of that fuel by 0.5 percent (through, for example, adoption of fuel-saving technologies and reduced use of energy-consuming products). The fiscal, health, and CO_2 impacts of these fuel and tax changes are then calculated. In addition, for coal it is assumed—based on a comparison indicating the fiscal incentives provided by corrective taxes for adopting

---

[6] For example, just because there might be a wide gap between the current and corrective tax for a particular fuel does not necessarily mean the fiscal and environmental benefits from reforming this fuel tax are larger than for other reform options, because these benefits also depend on fuel usage.

emissions control technologies are large relative to the costs of technology adoption—that implementing the corrective tax with appropriate crediting for mitigation during fuel combustion would lead to the adoption of control technologies at all coal plants that remain in operation.

These calculations leave aside a wide range of country-specific details, such as factors that might affect the price responsiveness of carbon-intensive fuels and the environmental and the fiscal implications of switching among fuels, but they are still useful in giving a broad sense of potential effects. Fiscal impacts are, however, overstated to the extent that compensation schemes, such as for low-income households, need to accompany tax reform.[7]

Fiscal Impacts

Despite the uncertainties surrounding these projections, a potentially large fiscal dividend from reforming fuel taxes clearly exists. The estimated dividend in Figure 6.11

Figure 6.11 Potential Revenue from Corrective Fuel Taxes, Selected Countries, 2010

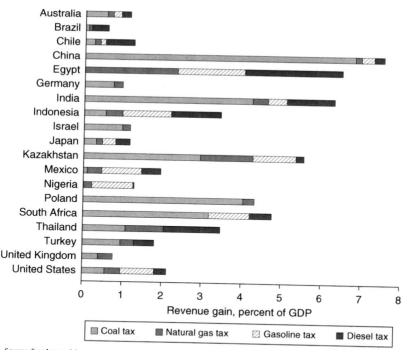

Source: See Annex 6.1.

Note: Figure shows revenue, expressed as a percentage of GDP, from corrective fuel taxes (allowing for behavioral responses to the tax) relative to current fuel tax revenues (which are often zero, and negative in cases where fuels are currently subsidized). In the few countries for which corrective taxes fall short of current taxes, the revenue potential is taken as zero.

---

[7] Dinan (forthcoming), for example, estimates that for the United States full compensation for the bottom income quintile for a carbon tax would offset 12 percent of the revenue.

is about 1 percent of GDP or more in all but two countries, and more than 3 percent in 8 countries, and 7.5 percent in China (which has a coal-intensive energy sector).[8] Even in Germany and the United Kingdom, where motor fuel taxes are relatively high, implementing corrective taxes on coal and natural gas results in estimated revenues of close to 1 percent of GDP. At a global level, revenue gains amount to 2.6 percent of world GDP.[9] These calculations do not take into account changes in value-added or similar taxes paid at the household level, though this additional revenue is relatively minor.

The composition of potential revenue also differs markedly across countries. For example, the corrective tax on coal is the dominant source of potential revenue in China, Germany, India, Israel, Kazakhstan, Poland, South Africa, and Turkey, while higher motor fuel taxes are the dominant source of potential revenue in Brazil, Chile, Egypt, Indonesia, Japan, Mexico, Nigeria, and the United States. Corrective taxes for natural gas also produce significant revenues in some cases (about 0.3 percent or more of GDP in 10 of the countries).

Health Impacts

Fuel tax reform can also dramatically reduce premature deaths from local air pollution, especially in countries that use large amounts of coal. Pollution-related deaths are reduced by more than half in nine of the countries shown in Figure 6.12. Reductions in emissions from coal combustion from the adoption of emissions control technologies and reduced use of coal are by far the main source of mortality reductions in most cases. At a global level, implementing corrective taxes reduces air pollution deaths by 63 percent.

Climate Impacts

Figure 6.13 illustrates another important benefit of fuel tax reform distinct from the health benefits of better local air quality: potentially large reductions in energy-related CO_2 emissions, expressed as annual percentage reductions in nationwide emissions for 2010. These reductions exceed 15 percent in all but two cases, and the greatest reduction in emissions is 34 percent in China.[10] At a global level, CO_2 reductions would amount to 23 percent.

---

[8] The fiscal dividend from the corrective coal tax in China, 6.8 percent of GDP, is evidently very large, but is based on the following assumptions: The corrective tax for local air pollution, at controlled emission rates, is $6.0/GJ which, along with the carbon charge of $3.3/GJ, increases the baseline coal price ($6.4/GJ) by about 145 percent. In turn, this reduces coal use by 36 percent to about 44 billion GJ. Multiplying this amount by the corrective tax ($9.3/GJ) gives revenue of about $405 billion, and dividing by 2010 GDP ($5,930 billion) gives the above figure.

[9] Clements and others (2013) estimate fossil fuel subsidies worldwide, including the implicit subsidy from the failure to charge for environmental side effects, to be almost 3 percent of global GDP. This estimate is based on a cruder assessment of external costs (using a simple extrapolation from several country case studies to the global level), which are lower than estimated here, however, the subsidy is assessed at current fuel consumption levels (rather than the lower levels that would occur with higher fuel prices).

[10] These estimated reductions implicitly reflect some combination of the main behavioral responses summarized in Figure 3.1.

Figure 6.12 Reduction in Pollution-Related Deaths from Corrective Fuel Taxes, Selected Countries, 2010

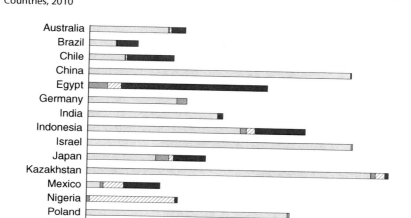

Source: See Annex 6.1.
Note: Figure shows percent reduction in premature deaths attributed to outdoor fossil fuel air pollution from implementing corrective fuel taxes relative to the current situation. In the few countries for which fossil-fuel corrective taxes fall short of current taxes, the tax is held fixed (so there are no reductions in deaths).

For all but five countries shown in Figure 6.13, coal—because of its high carbon intensity and the especially large increase in coal prices resulting from the corrective tax—accounts for 50 percent or more of the calculated total emissions reductions, and more than 85 percent of the reductions in China, India, Poland, and South Africa. In most countries, however, significant carbon reductions also occur from implementing corrective taxes on natural gas and motor fuels.

SUMMARY

Tax reforms can yield large fiscal dividends (2.6 percent of GDP worldwide), even in countries with high motor fuel taxes; significant reductions in global CO_2 emissions (23 percent); and especially from coal taxes, dramatic reductions in pollution-related deaths (63 percent).

Coal use in particular is highly and pervasively undercharged, not only for carbon emissions but also for the health costs of air pollution, though appropriate

Figure 6.13 Reduction in Energy-Related CO_2 Emissions from Corrective Fuel Taxes, Selected Countries, 2010

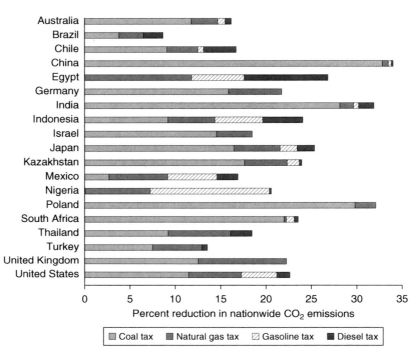

Source: See Annex 6.1.
Note: Figure shows reduction in nationwide energy-related CO_2 emissions (relative to 2010 levels) from implementing corrective fuel taxes relative to the current situation. In the few countries in which corrective taxes fall short of current taxes, the tax is held fixed (so there are no emission reductions).

charges for pollution differ considerably across countries. Heavy taxes on motor fuels are warranted in most developed and developing countries alike, but more to reflect the costs of traffic congestion and accidents rather than carbon emissions and pollution. For countries where motor fuel taxes are already high, the main opportunity for reform is to begin a progressive transition to kilometer-based charges to better manage congestion in particular. Although corrective charges for natural gas are small relative to those for coal (because of limited air pollution benefits), these charges can still generate significant revenue and CO_2 reductions. In short, much of the gains from energy price reform are in countries' own national interests.

ANNEX 6.1. ADDITIONAL DATA AND ASSUMPTIONS USED TO ESTIMATE THE IMPACTS OF FUEL TAX REFORM

Several additional pieces of information are needed to provide first-pass calculations of the impacts of fuel tax reform.

First, fuel prices faced by fuel users, primarily households and power plants, are needed by country, and are taken for 2010 from an IMF database compiled from multiple sources (see Clements and others, 2013, pp. 143–44).[11]

Second, the baseline quantities of the four fuels used in each country in 2010 are mostly taken from the database in Clements and others (2013), which was compiled from OECD and International Energy Agency (IEA) data. Diesel used by road vehicles was taken directly from IEA data. Fuel use reflects consumption by both households and firms—diesel fuel corresponds to that used by motor vehicles and natural gas to that used by power plants and other industrial sources as well as residential uses.

Third, the following, commonly used functional form is used to compute changes in fuel demand in response to changes in fuel prices:

$$Q_1 = \left(\frac{p_1}{p_0}\right)^\eta Q_0. \qquad (6.1)$$

In equation (6.1) Q and p denote, respectively, quantities and prices of a particular fuel, and subscripts 0 and 1 denote initial (current) values and values after adjusting the fuel tax rate to its corrective level. η denotes fuel price elasticity, or the percent change in fuel use per 1 percent increase in fuel price. Changes in fuel taxes are assumed to be fully reflected in the price paid by fuel users.[12]

The calculations simply assume that $\eta = -0.5$ for all fuels in all countries (i.e., each 1 percent increase in fuel price reduces fuel use by 0.5 percent). Numerous studies have estimated gasoline price elasticities for different countries, and the value assumed here roughly reflects a central value from the literature.[13] This assumption may, on average, overstate the price responsiveness of coal and natural gas somewhat (US EIA, 2012). If so, the fiscal impacts of the fuel tax reforms will be moderately understated, and the CO_2 and health impacts will be moderately overstated. It is difficult to make generalizations, however; for example, in countries with ample potential for renewables and nuclear fuels, coal and natural gas may have relatively large responsiveness, and the reverse in countries with little potential for these fuels. Also notable is that, to keep the calculations manageable, the fuel demand function in equation (6.1) is independent of

---

[11] For petroleum products, domestic prices to fuel users are taken from publicly available sources for OECD countries; for other countries they were provided by country authorities to the IMF, supplemented by survey data from Ebert and others (2009). Where fuel prices were unavailable they were imputed based on the observed pass-through of international fuel prices for that country. For coal and natural gas, for countries for which data are more sparse, Clements and others (2013) infer domestic prices from international prices, making an adjustment for transportation and distribution costs, and subtracting any fuel subsidies that have been quantified (for countries for which they are significant) by the OECD and the International Energy Agency.

[12] In reality, some portion might be passed backward into lower prices for fuel suppliers depending on the relative slope of fuel demand and supply curves, though this portion appears to be relatively modest (Bovenberg and Goulder, 2001).

[13] See various reviews cited in Parry, Walls, and Harrington (2007) and Sterner (2007).

other fuel prices. This lack of interaction will tend to overstate the impacts of full tax reform on fuel demand if fuels are substitutes (e.g., coal and natural gas in power generation, gasoline and diesel in passenger vehicles). In this regard, the revenue effects are moderately understated and the CO_2 and health effects overstated.

The fourth piece of information is current excise tax (or subsidy) rates. These rates are obtained from the database in Clements and others (2013), which also includes an estimate of fuel supply prices, based on a reference international fuel price (adjusted for transportation and distribution costs), applying to different regions. The excise tax (or subsidy) is the difference between the user price and the producer price, after netting out any applicable value-added taxes.

Changes in fuel use are calculated using equation (6.1); the difference between the new price and the initial fuel price is the difference between the corrective tax level and any existing excise tax (which is often zero, and sometimes negative, in which case the price increase exceeds the corrective tax).[14] Revenue from the corrective tax is simply the product of the excise tax and fuel use with this tax (Q_1). The change in revenue is this amount less the product of the initial tax and initial fuel use (Q_0). The revenue change exceeds revenue from the corrective tax if fuel is initially subsidized.

To calculate health effects, the corrective tax is assumed to provide incentives for all coal and natural gas plants to adopt emissions control technologies, in which case the relevant emissions factor is the one for representative plants in each country that already apply such technologies. This seems plausible, based on a quick comparison of the costs of installing and operating emissions control technologies, and the resulting tax credit that would ideally be provided if taxes were set at their corrective levels.[15] Mortality reductions are calculated by fuel use at the corrective tax times the weighted sum of SO_2, $PM_{2.5}$, and NO_x, where the weights are deaths per ton for these emissions, less initial fuel use, times the corresponding weighted sum of emissions. For the first product, controlled emissions factors are used. For the second, a weighted average of controlled and uncontrolled emissions factors is used.

Finally, CO_2 emissions reductions are computed based on the tax-induced fuel reductions and the fixed carbon emissions factors for the fuels.[16]

---

[14]The price increase is understated for coal and natural gas because it ignores additional costs at plants that apply emissions control technologies in response to the corrective tax.

[15]For example, using data supplied by Dallas Burtraw, the capital costs and discounted operating costs over a 10-year period for installing and using an SO_2 scrubber at a 500 megawatt coal plant in the United States would amount to about $300 million. And assuming the scrubber cuts emissions (that would otherwise be 42,000 tons a year) by 98 percent, the discounted tax savings, assuming a corrective tax of $17,000 per ton of SO_2, amount to about $5,500 million, or more than 18 times the technology cost.

[16]CO_2 emissions reductions are expressed against a baseline equal to emissions from the four fuels (which is slightly less than total energy-related emissions given that it excludes emissions from fuel products not considered here, principally petroleum products other than motor fuels).

ANNEX 6.2. CORRECTIVE ESTIMATES, THEIR IMPACTS, AND CURRENT TAXES, BY COUNTRY

The following tables summarize, country by country, the results related to corrective fuel taxes presented in the main text of this chapter. In particular Tables 6.2.1, 6.2.2, 6.2.3, and 6.2.4 provide, respectively, the estimated corrective taxes on coal, natural gas, gasoline, and motor diesel; the fiscal impacts of tax reform (revenue gains as a percentage of GDP); the percentage reduction in premature deaths and percentage reductions in CO_2 emissions from tax reform; and current excise taxes on fuel products.

ANNEX TABLE 6.2.1

Corrective Fuel Tax Estimates, All Countries, 2010

	Coal		Natural gas		Gasoline	Diesel
	$ per GJ		$ per GJ		$ per liter	
Country	average all plants	plants with controls	average all plants	ground level		
North America						
Canada	4.9	4.1	2.2	2.1	0.55	0.64
Mexico	3.9	3.6	2.0	2.1	0.31	0.40
United States	8.7	5.4	3.1	2.3	0.43	0.57
Central and South America						
Argentina	9.7	3.8	2.1	2.0	#na	#na
Barbados	#na	#na	3.2	#na	#na	#na
Bolivia	#na	#na	2.0	1.9	0.30	0.29
Brazil	4.4	3.6	2.0	2.0	0.39	0.45
Chile	4.1	3.6	2.0	2.0	0.56	0.68
Colombia	4.8	3.5	2.1	2.0	0.72	0.72
Costa Rica	#na	#na	#na	2.0	0.51	0.49
Cuba	#na	#na	2.3	2.0	#na	#na
Dominican Republic	5.7	3.7	#na	2.0	0.93	0.76
Ecuador	#na	#na	2.0	1.9	0.35	0.35
El Salvador	#na	#na	#na	1.9	0.55	0.42
Guatemala	4.0	3.4	#na	1.9	#na	#na
Honduras	#na	#na	#na	2.0	0.41	#na
Jamaica	#na	#na	#na	2.0	0.36	0.35
Nicaragua	#na	#na	2.0	#na	0.39	0.36
Panama	4.6	3.5	#na	2.0	0.47	0.48
Paraguay	#na	#na	#na	1.9	0.56	0.48
Peru	3.6	3.3	2.0	2.0	0.60	0.50
St. Vincent and the Grenadines	#na	#na	#na	#na	#na	0.16
Suriname	#na	#na	#na	1.9	0.22	0.38
Trinidad and Tobago	#na	#na	2.1	#na	#na	#na
Uruguay	#na	#na	2.2	2.0	0.65	0.54
Venezuela	#na	#na	2.1	2.0	0.48	0.57
Europe						
Albania	#na	#na	#na	#na	0.53	0.53
Austria	6.4	6.4	2.9	2.1	0.56	0.75
Belgium	20.4	20.4	3.0	2.1	0.80	0.90
Bosnia and Herzegovina	#na	#na	#na	2.0	0.37	0.47

(Continued)

ANNEX TABLE 6.2.1 (Continued)

Corrective Fuel Tax Estimates, All Countries, 2010

Country	Coal $ per GJ		Natural gas $ per GJ		Gasoline $ per liter	Diesel $ per liter
	average all plants	plants with controls	average all plants	ground level		
Bulgaria	57.0	9.7	#na	2.0	0.51	0.61
Croatia	36.1	11.0	4.6	2.1	0.46	0.66
Cyprus	#na	#na	#na	2.0	0.42	0.45
Czech Republic	18.6	18.6	3.2	2.1	0.53	0.69
Denmark	5.0	5.0	2.7	2.0	1.28	1.44
Finland	6.0	5.7	2.7	2.1	0.63	0.86
France	11.1	11.1	3.0	2.2	0.73	0.95
Germany	9.3	9.1	3.1	2.1	0.58	0.82
Greece	18.8	12.4	2.4	2.0	0.59	0.74
Hungary	25.5	15.5	3.7	2.1	0.46	0.54
Iceland	#na	#na	#na	#na	0.52	0.69
Ireland	6.9	5.1	2.3	2.0	0.61	0.72
Italy	5.8	5.8	2.6	2.1	0.53	0.75
Luxembourg	#na	#na	3.7	#na	0.86	0.91
Macedonia, FYR	35.3	7.9	4.1	2.0	#na	#na
Malta	#na	#na	#na	#na	0.47	0.58
Montenegro	34.2	6.5	#na	2.0	#na	#na
Netherlands	6.6	6.6	3.0	2.1	0.70	0.89
Norway	#na	#na	2.3	2.2	1.04	1.52
Poland	15.6	12.6	3.3	2.1	0.55	0.59
Portugal	6.9	5.3	2.3	2.0	0.52	0.64
Romania	39.2	12.8	3.9	2.0	0.59	0.63
Serbia	38.5	8.7	3.7	2.0	0.41	0.53
Slovak Republic	12.3	11.6	3.0	2.0	0.47	0.54
Slovenia	17.6	15.5	3.7	2.1	0.37	0.45
Spain	11.2	7.8	2.4	2.2	0.65	0.90
Sweden	5.1	5.1	2.4	2.1	0.63	0.85
Switzerland	#na	#na	3.5	2.1	0.94	1.13
Turkey	12.5	5.4	2.3	2.0	1.11	1.20
United Kingdom	14.7	12.7	3.0	2.1	0.60	0.77
Eurasia						
Armenia	#na	#na	2.3	2.0	0.30	0.20
Azerbaijan	#na	#na	2.5	2.0	0.64	0.69
Belarus	#na	#na	4.1	2.1	0.46	1.00
Estonia	#na	#na	2.7	2.0	0.24	0.37
Georgia	#na	#na	2.3	2.0	0.43	0.47
Kazakhstan	6.4	3.8	2.3	2.0	0.57	0.61
Kyrgyzstan	5.5	3.6	#na	1.9	0.31	0.26
Latvia	13.7	10.3	3.6	2.1	0.44	0.71
Lithuania	#na	#na	2.8	2.1	0.63	0.82
Russia	15.0	11.1	3.8	2.4	1.05	2.06
Tajikistan	#na	#na	#na	1.9	0.63	#na
Turkmenistan	#na	#na	2.3	2.0	#na	#na
Ukraine	32.9	10.1	3.6	2.0	0.39	0.50
Uzbekistan	6.3	4.1	2.1	1.9	#na	#na
Middle East						
Bahrain	#na	#na	2.3	2.0	0.31	0.36
Iran	#na	#na	2.2	2.0	0.60	0.59
Iraq	#na	#na	2.0	1.9	#na	#na

ANNEX TABLE 6.2.1

Corrective Fuel Tax Estimates, All Countries, 2010

	Coal		Natural gas		Gasoline	Diesel
	$ per GJ		$ per GJ		$ per liter	
Country	average all plants	plants with controls	average all plants	ground level		
Israel	22.0	9.2	2.7	#na	0.51	0.74
Jordan	#na	#na	2.0	2.0	0.34	0.31
Kuwait	#na	#na	#na	2.1	0.67	0.89
Lebanon	#na	#na	2.2	2.0	#na	#na
Oman	#na	#na	2.3	2.0	0.54	0.50
Qatar	#na	#na	2.8	2.0	#na	#na
Saudi Arabia	#na	#na	2.2	2.0	0.52	0.54
Syria	#na	#na	2.1	2.0	0.73	0.63
United Arab Emirates	#na	#na	2.2	2.0	#na	#na
Africa						
Algeria	#na	#na	2.1	2.0	#na	#na
Angola	#na	#na	2.0	2.0	#na	#na
Benin	#na	#na	#na	1.9	0.17	0.16
Botswana	4.1	3.4	2.0	#na	0.56	0.40
Burkina Faso	#na	#na	#na	1.9	0.12	0.13
Burundi	#na	#na	#na	1.9	0.13	0.13
Cabo Verde	#na	#na	#na	#na	0.83	0.51
Cameroon	#na	#na	2.0	1.9	0.15	0.17
Central African Republic	#na	#na	#na	1.9	0.82	0.49
Comoros	#na	#na	#na	#na	0.15	0.14
Congo, Rep. of	#na	#na	1.9	1.9	0.13	#na
Côte d'Ivoire	#na	#na	1.9	1.9	0.41	0.29
Egypt	#na	#na	2.1	2.0	0.38	0.35
Ethiopia	#na	#na	#na	1.9	0.41	0.27
Gambia, The	#na	#na	#na	1.9	0.14	0.14
Ghana	#na	#na	1.9	#na	0.28	0.23
Guinea-Bissau	#na	#na	#na	1.9	0.46	0.31
Kenya	#na	#na	1.9	#na	0.45	0.33
Liberia	#na	#na	#na	1.9	0.69	0.43
Libya	#na	#na	2.1	2.0	#na	#na
Madagascar	#na	#na	#na	1.9	0.21	0.18
Malawi	#na	#na	1.9	1.9	0.47	0.31
Mali	#na	#na	#na	1.9	0.38	0.26
Mauritius	3.6	3.3	#na	#na	0.57	0.37
Morocco	4.5	3.5	2.0	2.0	0.64	0.47
Mozambique	#na	#na	#na	1.9	0.44	0.29
Namibia	3.5	3.4	#na	1.9	#na	#na
Niger	#na	#na	#na	1.9	0.25	0.19
Nigeria	#na	#na	2.0	1.9	0.22	0.22
Rwanda	#na	#na	#na	1.9	0.31	0.23
São Tomé and Príncipe	#na	#na	#na	#na	0.24	0.19
Senegal	3.4	3.4	#na	1.9	0.18	0.18
Seychelles	#na	#na	#na	#na	0.52	0.34
Sierra Leone	#na	#na	#na	1.9	0.16	0.14
South Africa	4.6	3.6	2.0	#na	0.80	0.65
Sudan and South Suda	#na	#na	1.9	1.9	0.11	0.13
Swaziland	#na	#na	#na	#na	0.48	0.32

(Continued)

ANNEX TABLE 6.2.1 (Continued)
Corrective Fuel Tax Estimates, All Countries, 2010

	Coal		Natural gas		Gasoline	Diesel
	$ per GJ		$ per GJ		$ per liter	
Country	average all plants	plants with controls	average all plants	ground level		
Tanzania	#na	#na	1.9	1.9	0.51	0.33
Togo	#na	#na	1.9	**#na**	0.20	0.17
Tunisia	#na	#na	2.1	2.0	0.61	0.44
Uganda	#na	#na	#na	1.9	0.51	0.33
Zambia	#na	#na	#na	1.9	0.70	0.42
Zimbabwe	3.3	3.3	#na	1.9	0.22	0.20
Asia and Oceania						
Afghanistan	**#na**	**#na**	2.0	1.9	0.17	0.17
Australia	4.1	3.9	2.0	2.1	0.55	0.73
Bangladesh	8.3	4.2	2.3	2.0	0.45	0.43
Bhutan	**#na**	**#na**	#na	#na	0.96	0.58
Brunei	#na	#na	3.9	**#na**	0.29	0.25
Cambodia	**#na**	**#na**	#na	#na	0.73	0.52
China	15.0	9.2	3.2	2.5	0.55	0.51
Fiji	#na	#na	#na	#na	0.58	0.40
Hong Kong SAR	16.5	17.7	4.7	**#na**	**#na**	**#na**
India	8.7	8.7	2.2	2.0	0.78	0.54
Indonesia	5.7	4.0	2.3	2.0	0.32	0.42
Japan	5.5	5.1	3.1	2.5	1.13	1.44
Kiribati	#na	#na	#na	#na	0.18	0.17
Korea, Rep. of	8.1	7.7	4.1	2.3	0.98	1.20
Malaysia	5.3	4.7	2.3	2.0	0.55	0.58
Maldives	#na	#na	#na	#na	0.58	0.52
Mongolia	7.4	4.5	#na	#na	0.48	0.54
New Zealand	4.1	3.7	2.0	2.0	0.44	0.55
Pakistan	7.3	4.5	2.3	2.0	0.31	0.29
Papua New Guinea	#na	#na	**#na**	**#na**	0.34	0.26
Philippines	5.9	4.1	2.1	**#na**	0.25	0.32
Samoa	#na	#na	#na	#na	0.20	0.17
Singapore	**#na**	**#na**	2.9	2.9	1.29	2.18
Sri Lanka	5.7	3.7	#na	#na	0.63	0.42
Taiwan Province of China	6.0	6.3	4.4	**#na**	**#na**	**#na**
Thailand	12.7	5.6	2.3	**#na**	0.42	0.39
Vietnam	5.5	3.8	2.1	**#na**	0.46	0.43

Source: See Chapters 3 and 4.

Note: The table shows estimates of corrective taxes for coal and natural gas, reflecting combined damages from carbon and local pollution emissions; and motor fuels, reflecting combined damages from carbon and local pollution emissions, congestion, accidents, and, for diesel, road damage (from trucks). For coal, corrective taxes are shown averaged across all plants and across only plants with control technologies (when the two are the same, either all plants have control technologies or none do). Corrective motor fuel taxes are not reported when two or more components of the corrective tax cannot be estimated. Bold #na = data not available; black #na = fuel not used.

ANNEX TABLE 6.2.2
Fiscal Impacts of Tax Reform, All Countries, 2010 *(percent of GDP)*

Country	Coal tax revenue from corrective tax	Coal tax tax change	Natural gas tax revenue from corrective tax	Natural gas tax tax change	Gasoline tax revenue from corrective tax	Gasoline tax tax change	Diesel tax revenue from corrective tax	Diesel tax tax change
North America								
Canada	0.2	0.2	0.4	0.4	1.3	0.5	0.6	0.3
Mexico	0.1	0.1	0.4	0.4	1.1	1.0	0.5	0.5
United States	0.6	0.6	0.4	0.4	1.3	0.8	0.4	0.3
Central and South America								
Argentina	0.0	0.0	0.8	1.3	#na	#na	#na	#na
Barbados	#na	#na	0.0	0.0	#na	#na	#na	#na
Bolivia	#na	#na	0.9	0.9	1.4	1.3	1.5	2.6
Brazil	0.1	0.1	0.1	0.1	0.9	0.0	0.7	0.4
Chile	0.2	0.2	0.1	0.1	0.9	0.1	1.0	0.7
Colombia	0.1	0.1	0.2	0.2	1.0	0.1	0.9	0.6
Costa Rica	#na	#na	#na	#na	1.2	0.6	1.0	1.0
Cuba	#na	#na	#na	#na	#na	#na	#na	#na
Dominican Republic	0.1	0.1	#na	#na	1.7	1.0	0.7	0.6
Ecuador	#na	#na	0.1	0.1	1.0	2.7	0.5	2.1
El Salvador	#na	#na	#na	#na	1.2	1.2	0.8	0.9
Guatemala	0.1	0.1	#na	#na	#na	#na	#na	#na
Honduras	#na	#na	#na	#na	1.5	1.0	#na	#na
Jamaica	#na	#na	#na	#na	1.5	1.2	1.3	1.2
Nicaragua	#na	#na	#na	#na	1.2	0.7	#na	#na
Panama	0.0	0.0	#na	#na	1.1	1.3	0.9	1.3
Paraguay	#na	#na	#na	#na	1.3	0.5	2.5	2.2
Peru	0.1	0.1	0.3	0.3	0.6	0.1	1.2	0.9
St. Vincent and the Grenadines	#na	#na	#na	#na	#na	#na	#na	#na
Suriname	#na	#na	#na	#na	#na	0.0	#na	0.8
Trinidad and Tobago	#na	#na	7.1	7.1	#na	#na	#na	#na
Uruguay	#na	#na	0.0	0.0	0.8	0.1	1.0	0.2
Venezuela	#na	#na	0.5	0.8	0.3	2.8	0.1	0.7
Europe								
Albania	#na	#na	#na	#na	0.7	0.0	3.2	0.7
Austria	0.2	0.2	0.2	0.2	0.4	0.0	1.3	0.2
Belgium	0.3	0.3	0.4	0.4	0.3	0.0	1.5	0.4
Bosnia and Herzegovina	#na	#na	#na	#na	0.8	0.0	2.8	0.3
Bulgaria	4.4	4.4	#na	#na	0.9	0.0	2.2	0.3
Croatia	0.3	0.3	0.6	0.6	0.7	0.0	1.5	0.2
Cyprus	#na	#na	#na	#na	1.0	0.0	0.8	0.0
Czech Republic	4.1	4.1	0.4	0.4	0.7	0.0	1.5	0.0
Denmark	0.2	0.2	0.1	0.1	0.8	0.2	1.2	0.5
Finland	0.3	0.3	0.1	0.1	0.7	0.0	1.0	0.2
France	0.1	0.1	0.2	0.2	0.3	0.0	1.3	0.3
Germany	0.6	0.7	0.2	0.2	0.5	0.0	0.8	0.1
Greece	1.1	1.2	0.1	0.1	1.1	0.0	0.8	0.0
Hungary	0.9	0.9	0.9	0.9	0.7	0.0	1.3	0.0
Iceland	#na	#na	#na	#na	0.9	0.0	0.7	0.0
Ireland	0.1	0.2	0.2	0.2	0.6	0.0	1.0	0.0

(Continued)

ANNEX TABLE 6.2.2 (Continued)

Fiscal Impacts of Tax Reform, All Countries, 2010 *(percent of GDP)*

Country	Coal tax revenue from corrective tax	Coal tax tax change	Natural gas tax revenue from corrective tax	Natural gas tax tax change	Gasoline tax revenue from corrective tax	Gasoline tax tax change	Diesel tax revenue from corrective tax	Diesel tax tax change
Italy	0.1	0.1	0.3	0.3	0.4	0.0	1.0	0.1
Luxembourg	#na	#na	0.3	0.3	0.7	0.2	3.3	1.4
Macedonia, FYR	#na	#na	#na	#na	#na	#na	#na	#na
Malta	#na	#na	#na	#na	0.6	0.0	0.9	0.0
Montenegro	#na	#na	#na	#na	#na	#na	#na	#na
Netherlands	0.2	0.2	0.5	0.5	0.5	0.0	0.9	0.2
Norway	#na	#na	0.1	0.1	0.4	0.0	0.9	0.3
Poland	3.7	4.0	0.3	0.3	0.7	0.0	1.5	0.1
Portugal	0.1	0.1	0.2	0.2	0.5	0.0	1.5	0.1
Romania	1.7	1.7	0.8	0.8	0.7	0.0	1.3	0.2
Serbia	6.3	6.3	0.6	0.6	0.8	0.0	1.9	0.2
Slovak Republic	#na	#na	#na	#na	#na	#na	#na	#na
Slovenia	1.3	1.3	0.2	0.2	0.7	0.0	1.4	0.0
Spain	0.1	0.2	0.2	0.2	0.4	0.0	1.6	0.6
Sweden	0.1	0.1	0.0	0.0	0.7	0.0	0.8	0.0
Switzerland	#na	#na	0.1	0.1	0.7	0.2	0.5	0.2
Turkey	0.9	1.0	0.3	0.3	0.4	0.0	1.6	0.5
United Kingdom	0.4	0.4	0.4	0.4	0.6	0.0	1.0	0.0
Eurasia								
Armenia	#na	#na	1.1	1.1	0.8	0.2	0.3	0.2
Azerbaijan	#na	#na	1.3	1.8	1.2	1.0	0.5	0.6
Belarus	#na	#na	4.1	4.1	1.2	0.2	2.3	1.9
Estonia	#na	#na	0.3	0.3	0.5	0.0	1.1	0.0
Georgia	#na	#na	0.7	0.7	1.7	0.7	1.3	0.9
Kazakhstan	2.8	3.0	1.2	1.3	1.4	1.1	0.2	0.2
Kyrgyzstan	#na	#na	#na	#na	#na	#na	#na	#na
Latvia	0.1	0.1	0.7	0.7	0.8	0.0	2.3	0.5
Lithuania	#na	#na	0.6	0.6	0.7	0.0	2.2	0.8
Russia	2.2	2.2	3.0	3.9	1.9	1.9	1.0	1.1
Tajikistan	#na	#na	#na	#na	1.0	0.8	#na	#na
Turkmenistan	#na	#na	6.1	20.5	#na	#na	#na	#na
Ukraine	7.9	7.9	4.6	7.9	1.5	0.9	0.8	0.9
Uzbekistan	0.5	0.5	7.1	28.7	#na	#na	#na	#na
Middle East								
Bahrain	#na	#na	2.9	2.9	0.7	1.7	0.4	1.5
Iran	#na	#na	2.2	8.0	1.4	3.0	0.3	3.1
Iraq	#na	#na	0.2	0.4	#na	#na	#na	#na
Israel	1.0	1.0	0.2	0.2	0.8	0.0	0.6	0.0
Jordan	#na	#na	0.6	0.6	1.6	1.0	0.7	1.1
Kuwait	#na	#na	#na	#na	0.8	1.5	0.4	0.9
Lebanon	#na	#na	0.0	0.0	#na	#na	#na	#na
Oman	#na	#na	1.8	4.0	1.3	2.0	0.1	0.2
Qatar	#na	#na	2.0	3.5	#na	#na	#na	#na
Saudi Arabia	#na	#na	0.9	0.9	1.0	2.3	0.5	2.3
Syria	#na	#na	0.9	0.9	2.4	1.6	1.5	3.0
United Arab Emirates	#na	#na	1.4	4.8	#na	#na	#na	#na
Africa								
Algeria	#na	#na	1.0	6.2	#na	#na	#na	#na

ANNEX TABLE 6.2.2
Fiscal Impacts of Tax Reform, All Countries, 2010 (percent of GDP)

Country	Coal tax corrective tax	Coal tax revenue from tax change	Natural gas tax corrective tax	Natural gas tax revenue from tax change	Gasoline tax corrective tax	Gasoline tax revenue from tax change	Diesel tax corrective tax	Diesel tax revenue from tax change
Angola	#na	#na	0.0	0.0	#na	#na	#na	#na
Benin	#na	#na	#na	#na	2.1	0.2	1.3	0.4
Botswana	0.5	0.5	#na	#na	1.5	1.3	0.8	0.7
Burkina Faso	#na	#na	#na	#na	0.3	0.0	0.6	0.0
Burundi	#na	#na	#na	#na	0.3	0.0	0.7	0.0
Cabo Verde	#na	#na	#na	#na	1.7	0.2	1.9	0.7
Cameroon	#na	#na	0.1	0.1	0.3	0.0	0.5	0.1
Central African Republic	#na	#na	#na	#na	1.7	0.4	2.4	0.0
Comoros	#na	#na	#na	#na	#na	#na	#na	#na
Congo, Rep. of	#na	#na	#na	#na	#na	#na	#na	#na
Côte d'Ivoire	#na	#na	#na	#na	#na	#na	#na	#na
Egypt	#na	#na	1.3	2.3	0.7	1.7	0.5	2.5
Ethiopia	#na	#na	#na	#na	0.3	0.1	0.7	0.9
Gambia, The	#na	#na	#na	#na	0.4	0.0	0.7	0.0
Ghana	#na	#na	#na	#na	0.8	0.5	0.6	0.7
Guinea-Bissau	#na	#na	#na	#na	1.0	0.2	1.4	0.5
Kenya	#na	#na	#na	#na	1.1	0.4	1.0	1.0
Liberia	#na	#na	#na	#na	1.2	1.0	1.7	1.5
Libya	#na	#na	0.6	0.9	#na	#na	#na	#na
Madagascar	#na	#na	#na	#na	0.5	0.0	0.8	0.0
Malawi	#na	#na	#na	#na	1.1	0.0	1.6	0.0
Mali	#na	#na	#na	#na	0.9	0.0	1.2	0.3
Mauritius	0.5	0.5	#na	#na	1.4	0.2	1.2	0.6
Morocco	0.3	0.3	0.0	0.0	0.5	0.1	1.5	1.3
Mozambique	#na	#na	#na	#na	0.9	0.3	1.2	0.6
Namibia	0.1	0.1	#na	#na	#na	#na	#na	#na
Niger	#na	#na	#na	#na	0.5	0.1	0.9	0.0
Nigeria	#na	#na	0.2	0.2	0.5	1.0	0.0	0.0
Rwanda	#na	#na	#na	#na	0.8	0.0	1.4	0.0
São Tomé and Príncipe	#na	#na	#na	#na	#na	#na	#na	#na
Senegal	0.1	0.1	#na	#na	0.2	0.0	0.9	0.0
Seychelles	#na	#na	#na	#na	#na	#na	1.2	0.0
Sierra Leone	#na	#na	#na	#na	0.4	0.1	0.6	0.6
South Africa	3.2	3.2	0.1	0.1	1.7	1.0	1.0	0.6
Sudan and South Suda	#na	#na	#na	#na	#na	#na	#na	#na
Swaziland	#na	#na	#na	#na	0.9	0.5	1.2	0.6
Tanzania	#na	#na	0.2	0.2	0.7	0.2	1.3	0.2
Togo	#na	#na	#na	#na	1.6	0.0	0.7	0.1
Tunisia	#na	#na	0.7	0.7	0.7	0.5	1.5	1.2
Uganda	#na	#na	#na	#na	1.2	0.0	1.5	0.3
Zambia	#na	#na	#na	#na	0.9	0.1	0.9	0.0
Zimbabwe	2.7	2.7	#na	#na	0.7	0.0	0.7	0.1
Asia and Oceania								
Afghanistan	#na	#na	0.1	0.1	0.6	0.1	0.8	0.4
Australia	0.5	0.5	0.1	0.2	0.8	0.2	0.5	0.2

(Continued)

ANNEX TABLE 6.2.2 *(Continued)*

Fiscal Impacts of Tax Reform, All Countries, 2010 *(percent of GDP)*

Country	Coal tax revenue from corrective tax	Coal tax revenue from tax change	Natural gas tax revenue from corrective tax	Natural gas tax revenue from tax change	Gasoline tax revenue from corrective tax	Gasoline tax revenue from tax change	Diesel tax revenue from corrective tax	Diesel tax revenue from tax change
Bangladesh	0.1	0.1	1.2	2.9	0.2	0.1	0.3	0.7
Bhutan	#na	#na	#na	#na	1.4	1.0	1.8	2.3
Brunei	#na	#na	#na	#na	#na	#na	#na	#na
Cambodia	#na	#na	#na	#na	1.2	0.7	1.7	1.5
China	6.8	6.8	0.2	0.2	0.8	0.3	0.6	0.3
Fiji	#na	#na	#na	#na	#na	#na	#na	#na
Hong Kong SAR	#na	#na	#na	#na	#na	#na	#na	#na
India	3.1	3.1	0.3	0.4	0.8	0.5	1.0	1.2
Indonesia	0.5	0.5	0.4	0.4	0.9	1.2	0.5	1.3
Japan	0.3	0.3	0.2	0.2	1.1	0.3	0.5	0.4
Kiribati	#na	#na	#na	#na	#na	#na	#na	#na
Korea, Rep. of	#na	#na	#na	#na	#na	#na	#na	#na
Malaysia	0.8	0.8	0.9	1.3	2.0	2.1	0.9	1.2
Maldives	#na	#na	#na	#na	#na	#na	#na	#na
Mongolia	6.0	6.0	#na	#na	2.7	1.3	0.2	0.2
New Zealand	0.1	0.1	0.2	0.2	1.0	0.0	0.6	0.6
Pakistan	0.4	0.4	1.2	3.8	0.5	0.3	1.0	1.4
Papua New Guinea	#na	#na	#na	#na	0.4	#na	#na	#na
Philippines	0.5	0.5	0.1	0.1	0.5	0.1	0.6	0.9
Samoa	#na	#na	#na	#na	0.4	0.1	0.8	0.3
Singapore	#na	#na	0.3	0.3	0.5	0.3	1.1	0.9
Sri Lanka	0.0	0.0	#na	#na	0.9	0.6	1.1	2.0
Taiwan Province of China	#na	#na	#na	#na	#na	#na	#na	#na
Thailand	0.9	1.1	0.8	1.0	1.0	0.0	1.4	1.4
Vietnam	1.7	1.7	0.6	0.8	2.4	1.2	2.2	2.0

Source: See Annex 6.1.

Note: The table shows estimates of the revenue effect, as a percent of GDP, from imposing corrective taxes on different fuels, one column indicating the revenue from these taxes and the other the change in revenue from implementing the corrective tax compared with any revenue (or revenue losses) from existing taxes (or subsidies). Where current taxes exceed corrective taxes, revenue gains from tax reform are taken to be zero (for reasons discussed in Chapter 4, corrective motor fuel taxes may be understated, so lowering tax rates in these cases may not be warranted). To keep them manageable, the calculations do not account for the impact of taxes on one fuel affecting revenues from substitute fuels. Bold #na = data not available; black #na = fuel not used.

ANNEX TABLE 6.2.3

Health and Environmental Impacts of Tax Reform, All Countries, 2010

Country	Percent reduction in pollution deaths from				Percent reduction in nationwide energy-related CO_2 emissions			
	coal tax	natural gas tax	gasoline tax	diesel tax	coal tax	natural gas tax	gasoline tax	diesel tax
North America								
Canada	16.8	2.8	1.8	3.4	4.4	7.5	2.4	1.0
Mexico	3.4	0.8	5.2	9.2	2.7	6.5	5.4	2.3
United States	47.2	3.3	1.1	2.4	11.5	5.8	3.9	1.4
Central and South America								
Argentina	#na	#na	#na	#na	0.7	11.6	#na	#na
Barbados	#na	#na	#na	#na	#na	0.7	#na	#na
Bolivia	#na	1.5	1.5	20.0	#na	7.9	3.3	7.3
Brazil	6.8	0.2	0.0	5.4	3.8	2.7	0.0	2.1
Chile	9.0	0.1	0.4	11.8	9.1	3.4	0.6	3.6
Colombia	9.7	0.4	0.0	13.1	5.8	5.9	0.4	3.8
Costa Rica	#na	#na	1.4	13.3	#na	#na	5.0	7.6
Cuba	#na	#na	#na	#na	#na	1.4	#na	#na
Dominican Republic	22.7	#na	2.8	11.8	6.3	#na	6.4	4.3
Ecuador	#na	0.2	7.9	44.2	#na	1.2	23.6	21.3
El Salvador	#na	#na	3.1	13.8	#na	#na	10.8	7.8
Guatemala	#na	#na	#na	#na	4.3	#na	#na	#na
Honduras	#na	#na	#na	#na	#na	#na	5.0	#na
Jamaica	#na	#na	1.8	8.7	#na	#na	6.0	4.8
Nicaragua	#na	#na	0.9	#na	#na	#na	3.3	#na
Panama	4.8	#na	2.9	19.6	1.7	#na	10.1	11.2
Paraguay	#na	#na	0.4	12.3	#na	#na	2.0	9.6
Peru	1.5	0.4	0.3	9.1	2.3	6.5	0.6	3.7
St. Vincent and the Grenadines	#na	#na	#na	#na	#na	#na	#na	#na
Suriname	#na	#na	0.0	4.3	#na	#na	0.1	1.2
Trinidad and Tobago	#na	#na	#na	#na	#na	15.8	#na	#na
Uruguay	#na	0.1	0.0	0.0	#na	0.8	1.5	0.0
Venezuela	#na	1.7	29.0	50.5	#na	8.9	30.5	8.7
Europe								
Albania	#na	#na	0.0	0.0	#na	#na	0.1	0.0
Austria	9.2	4.0	0.0	0.0	8.6	7.3	0.0	0.0
Belgium	31.8	4.0	0.0	0.5	9.0	10.9	0.0	0.6
Bosnia and Herzegovina	#na	#na	0.0	0.0	#na	#na	0.0	0.0
Bulgaria	89.1	#na	0.0	0.0	17.5	#na	0.0	0.0
Croatia	51.1	5.3	0.0	0.0	7.5	12.0	0.0	0.0
Cyprus	#na	#na	0.0	0.0	#na	#na	0.0	0.0
Czech Republic	39.2	0.7	0.0	0.0	29.6	3.6	0.0	0.0
Denmark	13.6	4.5	0.0	4.3	12.0	5.4	1.4	2.8
Finland	17.1	2.2	0.0	0.7	12.5	4.4	0.0	0.3
France	13.8	3.3	0.0	1.7	7.2	7.8	0.0	1.2
Germany	22.3	2.6	0.0	0.0	15.9	5.9	0.0	0.0
Greece	48.0	0.2	0.0	0.0	9.7	2.3	0.0	0.0
Hungary	42.2	4.9	0.0	0.0	7.5	12.3	0.0	0.0
Iceland	#na	#na	0.0	0.0	#na	#na	0.0	0.0
Ireland	26.8	2.5	0.0	0.0	4.6	6.8	0.0	0.0

(Continued)

ANNEX TABLE 6.2.3 (Continued)
Health and Environmental Impacts of Tax Reform, All Countries, 2010

Country	Percent reduction in pollution deaths from				Percent reduction in nationwide energy-related CO_2 emissions			
	coal tax	natural gas tax	gasoline tax	diesel tax	coal tax	natural gas tax	gasoline tax	diesel tax
Italy	6.2	4.6	0.0	0.0	5.9	9.8	0.0	0.0
Luxembourg	#na	#na	#na	#na	#na	5.7	0.6	4.4
Macedonia, FYR	#na	#na	#na	#na	19.1	0.7	#na	#na
Malta	#na	#na	#na	#na	#na	#na	0.0	0.0
Montenegro	#na	#na	#na	#na	19.6	#na	#na	#na
Netherlands	9.5	9.3	0.0	0.4	7.0	12.7	0.0	0.2
Norway	#na	1.6	0.0	7.7	#na	9.0	0.1	2.2
Poland	50.7	0.6	0.0	0.0	29.8	2.3	0.0	0.0
Portugal	19.7	1.3	0.0	0.0	5.5	5.2	0.0	0.0
Romania	71.2	1.9	0.0	0.0	10.4	9.5	0.0	0.0
Serbia	84.1	0.3	0.0	0.0	22.6	2.6	0.0	0.0
Slovak Republic	33.2	2.9	#na	#na	17.0	7.9	#na	#na
Slovenia	36.4	1.6	0.0	0.0	15.4	3.8	0.0	0.0
Spain	17.3	1.3	0.0	3.9	5.1	6.9	0.0	2.3
Sweden	4.4	0.6	0.0	0.0	6.5	1.8	0.0	0.0
Switzerland	#na	8.0	0.3	3.3	#na	4.0	1.9	1.0
Turkey	72.9	0.6	0.0	0.3	7.5	5.4	0.1	0.6
United Kingdom	37.5	3.8	0.0	0.0	12.5	9.7	0.0	0.0
Eurasia								
Armenia	#na	#na	#na	#na	#na	5.1	0.2	0.1
Azerbaijan	#na	6.4	9.6	11.1	#na	11.7	3.1	1.8
Belarus	#na	14.3	0.2	11.7	#na	21.9	0.3	2.7
Estonia	#na	2.8	0.0	0.0	#na	1.3	0.0	0.0
Georgia	#na	1.6	2.9	5.8	#na	8.8	2.2	2.1
Kazakhstan	68.6	1.1	2.4	1.0	18.1	4.7	1.3	0.3
Kyrgyzstan	#na	#na	#na	#na	6.5	#na	#na	#na
Latvia	6.4	6.0	0.0	0.0	2.4	12.1	0.0	0.0
Lithuania	#na	3.5	0.0	4.9	#na	11.5	0.0	1.7
Russia	25.0	5.7	5.9	9.5	11.5	15.0	2.7	1.4
Tajikistan	#na	#na	#na	#na	#na	#na	2.8	#na
Turkmenistan	#na	#na	#na	#na	#na	15.4	#na	#na
Ukraine	75.3	1.9	0.0	0.2	18.7	10.7	0.5	0.4
Uzbekistan	#na	#na	#na	#na	0.9	14.8	#na	#na
Middle East								
Bahrain	#na	10.9	3.1	17.2	#na	14.7	4.4	3.8
Iran	#na	#na	10.9	3.1	#na	12.0	8.4	12.3
Iraq	#na	#na	#na	#na	#na	3.8	#na	#na
Israel	66.4	0.4	0.0	0.0	14.5	3.9	0.0	0.0
Jordan	#na	3.1	1.8	13.2	#na	8.0	3.4	3.9
Kuwait	#na	#na	13.2	48.7	#na	#na	11.4	6.6
Lebanon	#na	#na	#na	#na	#na	1.3	#na	#na
Oman	#na	13.5	7.0	3.6	#na	14.8	7.1	0.6
Qatar	#na	#na	#na	#na	#na	17.8	#na	#na
Saudi Arabia	#na	2.3	7.8	55.5	#na	10.0	13.0	14.5
Syria	#na	3.0	1.8	32.1	#na	9.6	3.6	9.9
United Arab Emirates	#na	#na	#na	#na	#na	13.9	#na	#na
Africa								
Algeria	#na	#na	#na	#na	#na	10.4	#na	#na

ANNEX TABLE 6.2.3
Health and Environmental Impacts of Tax Reform, All Countries, 2010

Country	Percent reduction in pollution deaths from				Percent reduction in nationwide energy-related CO_2 emissions			
	coal tax	natural gas tax	gasoline tax	diesel tax	coal tax	natural gas tax	gasoline tax	diesel tax
Angola	#na	#na	#na	#na	#na	2.4	#na	#na
Benin	#na	#na	0.0	0.0	#na	#na	0.6	0.0
Botswana	38.6	#na	3.9	4.1	12.7	#na	5.0	2.5
Burkina Faso	#na	#na	0.0	0.0	#na	#na	0.0	0.0
Burundi	#na	#na	0.0	0.0	#na	#na	0.0	0.0
Cabo Verde	#na	#na	#na	#na	#na	#na	0.6	1.2
Cameroon	#na	0.1	0.0	0.0	#na	2.9	0.0	0.0
Central African Republic	#na	#na	1.0	0.0	#na	#na	1.9	0.0
Comoros	#na	#na	#na	#na	#na	#na	#na	#na
Congo, Rep. of	#na	#na	#na	#na	#na	4.5	#na	#na
Côte d'Ivoire	#na	1.5	#na	#na	#na	7.4	#na	#na
Egypt	#na	4.8	3.5	36.8	#na	11.8	5.8	9.2
Ethiopia	#na	#na	0.9	11.5	#na	#na	1.7	10.5
Gambia, The	#na	#na	0.0	0.0	#na	#na	0.0	0.0
Ghana	#na	#na	3.1	6.7	#na	#na	4.9	5.0
Guinea-Bissau	#na	#na	0.0	0.0	#na	#na	0.3	0.2
Kenya	#na	#na	1.5	8.6	#na	#na	2.3	6.3
Liberia	#na	#na	3.5	9.6	#na	#na	2.6	3.4
Libya	#na	#na	#na	#na	#na	8.3	#na	#na
Madagascar	#na	#na	0.0	0.0	#na	#na	0.0	0.0
Malawi	#na	#na	0.0	0.0	#na	#na	0.0	0.0
Mali	#na	#na	0.0	0.0	#na	#na	0.0	0.0
Mauritius	#na	#na	#na	#na	8.6	#na	0.4	0.7
Morocco	27.4	0.1	0.4	8.8	11.3	0.9	0.5	5.8
Mozambique	#na	#na	#na	#na	#na	#na	1.5	1.0
Namibia	#na	#na	#na	#na	3.3	#na	#na	#na
Niger	#na	#na	0.0	0.0	#na	#na	0.7	0.0
Nigeria	#na	0.8	21.4	0.7	#na	7.2	13.1	0.2
Rwanda	#na	#na	0.0	0.0	#na	#na	0.0	0.0
São Tomé and Príncipe	#na	#na	#na	#na	#na	#na	#na	#na
Senegal	9.4	#na	0.0	0.0	5.3	#na	0.0	0.0
Seychelles	#na	#na	#na	#na	#na	#na	#na	0.0
Sierra Leone	#na	#na	0.0	1.7	#na	#na	0.3	0.4
South Africa	67.5	0.0	0.4	1.3	22.2	0.2	0.9	0.4
Sudan and South Sudan	#na	#na	#na	#na	#na	#na	#na	#na
Swaziland	#na	#na	#na	#na	#na	#na	2.2	1.3
Tanzania	#na	0.3	0.7	0.0	#na	4.7	0.9	0.0
Togo	#na	#na	0.0	0.0	#na	#na	0.0	0.0
Tunisia	#na	2.5	2.2	9.1	#na	10.4	1.6	3.6
Uganda	#na	#na	0.0	0.0	#na	#na	0.0	0.0
Zambia	#na	#na	0.0	0.0	#na	#na	1.4	0.0
Zimbabwe	61.1	#na	0.0	0.0	19.8	#na	0.0	0.0
Asia and Oceania								
Afghanistan	#na	0.1	0.0	0.0	#na	0.7	0.2	0.0
Australia	19.8	0.3	0.5	3.6	11.8	2.9	0.8	0.7

(Continued)

ANNEX TABLE 6.2.3 *(Continued)*

Health and Environmental Impacts of Tax Reform, All Countries, 2010

Country	Percent reduction in pollution deaths from				Percent reduction in nationwide energy-related CO_2 emissions			
	coal tax	natural gas tax	gasoline tax	diesel tax	coal tax	natural gas tax	gasoline tax	diesel tax
Bangladesh	18.5	7.2	0.6	9.5	1.5	15.0	0.3	2.2
Bhutan	#na	#na	**#na**	**#na**	#na	#na	4.7	12.4
Brunei	#na	**#na**	**#na**	**#na**	#na	20.4	**#na**	**#na**
Cambodia	**#na**	#na	2.6	13.3	**#na**	#na	2.0	4.5
China	65.9	0.1	0.0	0.1	32.8	0.7	0.3	0.2
Fiji	#na	#na	**#na**	**#na**	#na	#na	**#na**	**#na**
Hong Kong SAR	#na	**#na**	**#na**	**#na**	15.5	**#na**	**#na**	**#na**
India	63.1	0.2	0.2	1.0	22.0	1.5	0.6	1.7
Indonesia	38.4	1.5	2.2	12.6	9.2	5.2	5.3	4.4
Japan	17.2	3.3	1.2	8.1	16.5	5.1	1.9	1.9
Kiribati	#na	#na	**#na**	**#na**	#na	#na	**#na**	**#na**
Korea, Rep. of	3.7	3.9	0.1	**#na**	27.2	5.3	0.5	**#na**
Malaysia	20.0	2.8	2.4	11.6	9.3	7.0	5.1	3.0
Maldives	#na	#na	**#na**	**#na**	#na	#na	**#na**	**#na**
Mongolia	71.4	**#na**	0.3	0.2	13.5	#na	0.9	0.1
New Zealand	14.1	0.4	**#na**	12.7	4.9	5.1	0.0	3.4
Pakistan	31.1	4.3	0.5	4.6	2.3	10.3	0.6	3.0
Papua New Guinea	**#na**	**#na**	**#na**	**#na**	#na	**#na**	**#na**	**#na**
Philippines	51.3	0.2	0.1	5.0	12.9	1.9	0.6	3.3
Samoa	#na	#na	**#na**	**#na**	#na	#na	**#na**	**#na**
Singapore	**#na**	2.6	0.9	34.9	**#na**	1.6	0.3	1.0
Sri Lanka	6.7	**#na**	2.6	21.7	#na	**#na**	2.9	11.7
Taiwan Province of China	#na	**#na**	**#na**	**#na**	20.2	3.4	**#na**	**#na**
Thailand	66.3	1.1	0.0	1.7	9.2	6.9	0.0	2.4
Vietnam	44.2	0.3	1.6	5.1	12.4	2.9	1.7	2.6

Source: See Annex 6.2.

Note: The table shows the percent reduction in countries' nationwide deaths from air pollution and the percent reduction in nationwide CO_2 emissions from implementing corrective taxes on each of the fuels (ignoring impacts on use of other fuels). The reduction in CO_2 emissions comes from the sorts of behavioral responses summarized in Figure 3.1 while the reduction in premature deaths comes from similar responses and the adoption of emissions control equipment at power plants. In cases where current taxes exceed corrective taxes, potential health and CO_2 reductions from tax reform are indicated by 0. Bold #na = data not available; black #na = fuel not used.

ANNEX TABLE 6.2.4
Estimates of Current Fuel Excise Taxes, All Countries, 2010

	Current excise taxes			
Country	$ per gigajoule coal	$ per gigajoule natural gas	$ per liter gasoline	$ per liter diesel
North America				
Canada	0.0	−0.2	0.36	0.42
Mexico	0.0	0.0	0.10	0.10
United States	0.0	−0.1	0.13	0.14
Central and South America				
Argentina	0.0	−1.3	0.33	0.39
Barbados	0.0	0.0	0.42	0.28
Bolivia	0.0	0.0	0.07	−0.12
Brazil	0.0	0.0	0.75	0.28
Chile	0.0	0.0	0.60	0.26
Colombia	0.0	0.0	0.78	0.29
Costa Rica	0.0	0.0	0.31	0.11
Cuba	#na	#na	**#na**	**#na**
Dominican Republic	0.0	0.0	0.40	0.17
Ecuador	0.0	0.0	−0.32	−0.45
El Salvador	0.0	0.0	0.09	0.03
Guatemala	0.0	0.0	0.13	0.04
Honduras	0.0	0.0	0.21	0.06
Jamaica	0.0	0.0	0.15	0.12
Nicaragua	0.0	0.0	0.26	0.00
Panama	0.0	0.0	0.02	−0.09
Paraguay	0.0	0.0	0.45	0.15
Peru	0.0	0.0	0.58	0.24
St. Vincent and the Grenadines	#na	#na	**#na**	**#na**
Suriname	0.0	0.0	0.31	0.26
Trinidad and Tobago	0.0	0.0	0.00	0.00
Uruguay	0.0	0.0	0.66	0.58
Venezuela	0.0	−1.3	−0.60	−0.65
Europe				
Albania	0.0	0.0	0.63	0.54
Austria	0.0	0.0	0.93	0.78
Belgium	0.0	0.0	1.18	0.83
Bosnia and Herzegovina	0.0	0.0	0.59	0.56
Bulgaria	0.0	0.0	0.71	0.65
Croatia	0.0	0.0	0.86	0.73
Cyprus	#na	0.0	0.69	0.66
Czech Republic	0.0	0.0	0.99	0.89
Denmark	0.0	0.0	1.16	0.87
Finland	0.0	0.0	1.22	0.80
France	0.0	0.0	1.13	0.84
Germany	−1.3	0.0	1.20	0.92
Greece	−0.1	0.0	1.29	0.89
Hungary	0.0	0.0	0.95	0.82
Iceland	#na	0.0	0.88	0.85
Ireland	−3.3	0.0	1.05	0.90
Italy	0.0	0.0	1.08	0.86
Luxembourg	#na	0.0	0.84	0.60
Macedonia, FYR	#na	#na	**#na**	**#na**
Malta	#na	0.0	0.87	0.73
Montenegro	#na	#na	**#na**	**#na**
Netherlands	0.0	0.0	1.31	0.84

(Continued)

ANNEX TABLE 6.2.4 (*Continued*)

Estimates of Current Fuel Excise Taxes, All Countries, 2010

	Current excise taxes			
Country	$ per gigajoule coal	$ per gigajoule natural gas	$ per liter gasoline	$ per liter diesel
Norway	0.0	−0.2	1.34	1.11
Poland	−0.6	0.0	0.85	0.71
Portugal	−0.1	0.0	1.12	0.77
Romania	0.0	0.0	0.76	0.69
Serbia	0.0	0.0	0.67	0.62
Slovak Republic	#na	#na	#na	#na
Slovenia	−0.4	0.0	0.96	0.86
Spain	−2.0	0.0	0.85	0.70
Sweden	0.0	0.0	1.19	1.01
Switzerland	#na	0.0	0.87	0.92
Turkey	−0.5	0.0	1.37	0.97
United Kingdom	0.0	−0.1	1.20	1.21
Eurasia				
Armenia	#na	0.0	0.28	0.14
Azerbaijan	#na	−0.9	0.09	−0.05
Belarus	0.0	0.0	0.45	0.20
Estonia	0.0	0.0	0.83	0.79
Georgia	0.0	0.0	0.30	0.23
Kazakhstan	−0.2	−0.2	0.12	−0.08
Kyrgyzstan	#na	#na	#na	#na
Latvia	0.0	0.0	0.76	0.71
Lithuania	0.0	0.0	0.87	0.62
Russia	0.0	−0.9	0.02	0.02
Tajikistan	0.0	0.0	0.13	0.10
Turkmenistan	0.0	−4.4	−0.34	−0.38
Ukraine	0.0	−1.9	0.18	0.06
Uzbekistan	0.0	−5.3	0.17	0.08
Middle East				
Bahrain	0.0	0.0	−0.30	−0.38
Iran	0.0	−4.8	−0.37	−0.55
Iraq	0.0	−1.6	−0.18	−0.23
Israel	0.0	−0.2	0.93	0.94
Jordan	0.0	0.0	0.14	−0.09
Kuwait	0.0	−1.7	−0.32	−0.38
Lebanon	0.0	0.0	0.36	−0.06
Oman	0.0	−2.2	−0.24	−0.19
Qatar	0.0	−1.6	−0.33	−0.38
Saudi Arabia	0.0	0.0	−0.38	−0.50
Syria	0.0	0.0	0.27	−0.34
United Arab Emirates	0.0	−4.6	−0.20	0.03
Africa				
Algeria	0.0	−8.6	−0.24	−0.39
Angola	0.0	0.0	−0.12	−0.31
Benin	0.0	0.0	0.20	0.22
Botswana	0.0	0.0	0.10	0.11
Burkina Faso	0.0	0.0	0.57	0.38
Burundi	0.0	0.0	0.63	0.60
Cabo Verde	0.0	0.0	0.99	0.44
Cameroon	0.0	0.0	0.37	0.24
Central African Republic	0.0	0.0	0.81	0.76
Comoros	0.0	0.0	0.00	0.00

ANNEX TABLE 6.2.4
Estimates of Current Fuel Excise Taxes, All Countries, 2010

	Current excise taxes			
Country	$ per gigajoule coal	$ per gigajoule natural gas	$ per liter gasoline	$ per liter diesel
Congo, Rep. of	#na	#na	#na	#na
Côte d'Ivoire	#na	#na	#na	#na
Egypt	#na	−1.4	−0.42	−0.57
Ethiopia	0.0	0.0	0.23	0.00
Gambia, The	0.0	0.0	0.56	0.35
Ghana	0.0	0.0	0.11	0.05
Guinea-Bissau	0.0	0.0	0.52	0.29
Kenya	0.0	0.0	0.38	0.07
Liberia	0.0	0.0	0.17	0.14
Libya	0.0	−1.1	−0.41	−0.46
Madagascar	0.0	0.0	0.57	0.31
Malawi	0.0	0.0	0.88	0.68
Mali	0.0	0.0	0.51	0.30
Mauritius	0.0	0.0	0.62	0.29
Morocco	0.0	0.0	0.61	0.14
Mozambique	0.0	0.0	0.38	0.25
Namibia	0.0	0.0	0.23	0.23
Niger	0.0	0.0	0.24	0.30
Nigeria	0.0	0.0	−0.19	0.11
Rwanda	0.0	0.0	1.00	0.96
São Tomé and Príncipe	#na	#na	#na	#na
Senegal	0.0	0.0	0.59	0.32
Seychelles	0.0	0.0	0.00	0.55
Sierra Leone	0.0	0.0	0.14	0.10
South Africa	0.0	0.0	0.35	0.34
Sudan and South Suda	#na	#na	#na	#na
Swaziland	0.0	0.0	0.24	0.24
Tanzania	0.0	0.0	0.48	0.40
Togo	0.0	0.0	0.31	0.26
Tunisia	0.0	0.0	0.22	0.16
Uganda	0.0	0.0	0.73	0.39
Zambia	0.0	0.0	0.76	0.59
Zimbabwe	0.0	0.0	0.46	0.29
Asia and Oceania				
Afghanistan	#na	0.0	0.20	0.18
Australia	0.0	−0.1	0.49	0.49
Bangladesh	0.0	−2.6	0.26	−0.23
Bhutan	#na	0.0	0.25	−0.04
Brunei	#na	#na	#na	#na
Cambodia	#na	0.0	0.32	0.12
China	0.0	0.0	0.39	0.37
Fiji	#na	0.0	0.00	0.00
Hong Kong SAR	#na	#na	#na	#na
India	0.0	−1.0	0.36	−0.04
Indonesia	0.0	0.0	−0.13	−0.35
Japan	0.0	0.0	0.75	0.46
Kiribati	#na	0.0	0.00	0.00
Korea, Rep. of	#na	#na	0.85	#na
Malaysia	0.0	−0.8	−0.04	−0.10
Maldives	#na	0.0	0.00	0.00
Mongolia	0.0	0.0	0.28	0.18

(Continued)

ANNEX TABLE 6.2.4 *(Continued)*

Estimates of Current Fuel Excise Taxes, All Countries, 2010

	Current excise taxes			
Country	$ per gigajoule coal	$ per gigajoule natural gas	$ per liter gasoline	$ per liter diesel
New Zealand	0.0	0.0	0.63	0.13
Pakistan	0.0	−4.1	0.17	−0.03
Papua New Guinea	#na	0.0	0.00	0.00
Philippines	0.0	0.0	0.22	−0.02
Samoa	#na	0.0	0.20	0.20
Singapore	#na	0.0	0.69	0.31
Sri Lanka	0.0	0.0	0.21	−0.20
Taiwan Province of China	#na	#na	**#na**	**#na**
Thailand	−0.9	−0.3	0.59	0.09
Vietnam	0.0	−0.7	0.25	0.11

Source: See Annex 6.1.

Note: The table shows estimates of the current excise tax (or subsidy) for each fuel as measured in Clements and others (2013) using the price-gap approach (e.g., comparing differences between domestic and international fuel prices) consistently applied across countries. These estimates will differ from authorities' assessments of tax rates based on their own country-specific data. Bold #na = data not available; black #na = fuel not used.

REFERENCES

Bovenberg, Lans A., and Lawrence H. Goulder, 2001, "Neutralizing the Adverse Impacts of CO_2 Abatement Policies: What Does It Cost?" in *Behavioral and Distributional Effects of Environmental Policy*, edited by C. Carraro and G. Metcalf (Chicago, Illinois: University of Chicago Press).

Clements, Benedict, David Coady, Stefania Fabrizio, Sanjeev Gupta, Trevor Alleyene, and Carlo Sdralevich, eds., 2013, *Energy Subsidy Reform: Lessons and Implications* (Washington: International Monetary Fund).

Cropper, Maureen, Shama Gamkhar, Kabir Malik, Alex Limonov, and Ian Partridge, 2012, "The Health Effects of Coal Electricity Generation in India," Discussion Paper No. 12–15 (Washington: Resources for the Future).

Dinan, Terry, forthcoming, "Offsetting a Carbon Tax's Burden on Low-Income Households," in *Implementing a U.S. Carbon Tax: Challenges and Debates*, edited by I. Parry, A. Morris, and R. Williams (Washington: International Monetary Fund).

Ebert, Sebastian, Gerhard P. Metschies, Dominik Schmid, and Armin Wagner, 2009, *International Fuel Prices 2009* (Eschborn, Germany: Deutsche Gesellschaft für Internationale Zusammenarbeit).

International Monetary Fund (IMF), 2013, World Economic Outlook Database (Washington: International Monetary Fund). http://www.imf.org/external/pubs/ft/weo/2013/01/weodata/index.aspx.

Parry, Ian W.H., Margaret Walls, and Winston Harrington, 2007, "Automobile Externalities and Policies," *Journal of Economic Literature*, Vol. 45, pp. 374–400.

Sterner, Thomas, 2007, "Fuel Taxes: An Important Instrument for Climate Policy," *Energy Policy*, Vol. 35, No. 3, pp. 3194–202.

United States Energy Information Administration (US EIA), 2012, *Fuel Competition in Power Generation and Elasticities of Substitution* (Washington: Energy Information Administration, US Department of Energy).

World Health Organization (WHO), 2013, *Global Health Observatory Data Repository* (Geneva: World Health Organization).

CHAPTER 7

Concluding Thoughts

Encouraging environmentally sustainable growth is a problem faced by all countries. The beauty of fiscal instruments such as environmental taxes or tax-like instruments is that (albeit with some important caveats about base targeting, exploiting fiscal opportunities, and use of complementary instruments) they can achieve an efficient balance between environmental and economic concerns—*if* they are set to reflect environmental damage.

This volume shows how environmental damage can be measured for different countries, focusing on damage related to the use of fossil fuels, and how this information can be used to put into practice the principle of "getting prices right."

This exercise involves a multitude of caveats. From an analytical perspective, legitimate questions arise about data reliability and the methods used to quantify environmental damage. However, the accuracy of environmental damage assessments has been improving dramatically, and will continue to do so, not least because of modeling efforts elsewhere.[1] And although legitimate disagreements will remain, for example, about the appropriate values for carbon emissions and pollution- and accident-related premature deaths, spreadsheets accompanying this report can help discipline this debate by clarifying how alternative assumptions affect the economically efficient system of energy taxes.

Implementing efficient energy tax systems is highly challenging, especially because of resistance to higher energy prices. However, having some sense of the direction in which policy should ideally be headed from an economic perspective is very useful. It provides a benchmark against which other, perhaps politically easier, options can be judged, giving policymakers a better sense of the trade-offs they face, for example, in the environmental effectiveness and the revenue from well-designed tax reforms versus the weaker environmental effectiveness and lack of revenue from regulatory approaches. Moreover, the efficient scale of other, nonfiscal instruments, such as standards for energy efficiency and renewables, might be evaluated by comparing the incremental costs they impose per ton of emissions with the environmental damages per ton estimates here.

The hope is that this book will help countries move forward with policy reforms, as well as stimulate further analytical work and the data collection needed to improve the accuracy of country-by-country damage assessments, all of which

---

[1] For example, the Intergovernmental Panel on Climate Change on climate impacts, and the Global Burden of Disease project, the International Institute for Applied Systems Analysis, and the Climate and Clean Air Coalition on air pollution damage.

promote more informed policy decisions. The findings suggest large and pervasive disparities between efficient fuel taxes and current practice in developed and developing countries alike—with much at stake for health, environmental, and fiscal outcomes. There is much to be done in getting energy prices right; the aim of this volume is to provide the tools and evidence to help achieve this in practice.

Glossary

Air emissions regulations. Requirements for the use of emissions control technologies or standards for allowable emission rates (e.g., per kWh averaged over a generator's plants).

Air quality model. Computational models that link emissions from different sources to ambient pollution concentrations in nearby and more distant regions, accounting for meteorological and other factors that influence pollution formation.

Area licensing scheme. A scheme charging motorists for driving in a restricted area with high congestion.

Biofuels. Fuels that contain energy from recent production of carbon in living organisms such as plants and algae.

Border adjustments. The use of charges on the embodied pollution in imported products to alleviate concerns about the impact of environmental taxes on the international competitiveness of domestic firms.

Breathing rate (BR). The rate at which a given amount of outdoor air pollution is inhaled by the average person.

Carbon capture and storage (CCS). Technologies to separate carbon dioxide emissions during fuel combustion at, for example, coal plants, transport it to a storage site, and deposit it in an underground geological formation (e.g., depleted gas fields) to prevent its release into the atmosphere.

Carbon dioxide (CO_2). The predominant greenhouse gas. To convert tons of carbon dioxide into tons of carbon, divide by 3.67. To convert a price per ton of carbon dioxide into a price per ton of carbon, multiply by 3.67.

Carbon dioxide (CO_2) equivalent. The global warming potential of a greenhouse gas over its atmospheric lifespan (or over a long period) expressed as the amount of carbon dioxide that would yield the same amount of warming.

Carbon Monitoring for Action (CARMA). A database used for obtaining the geographical location of coal and natural gas power plants across different countries.

Carbon tax. A tax imposed on carbon dioxide emissions released largely through the combustion of carbon-based fossil fuels.

Common but differentiated responsibilities. A principle of the United Nations Framework Convention on Climate Change calling for developed countries to bear a disproportionately larger burden of mitigation costs (e.g., by funding emissions reduction projects in developing countries), given that they are relatively wealthy and contributed most to historical atmospheric greenhouse gas accumulations.

Concentration response function. The relationship between ambient pollution concentrations and elevated risks of various fatal diseases for populations exposed to the pollution.

Corrective tax. A charge levied on a source of environmental harm and that is set at a level to reflect, or correct for, environmental damage.

Cost-effective environmental policy. A policy that achieves a given level of environmental protection at minimum economic cost. This requires (1) pricing policies to equate incremental mitigation costs across different sources of an environmental harm and (2) that the revenue potential from pricing policies be realized and revenues be used productively (e.g., to lower other taxes that distort economic activity).

Credit trading. In emissions trading systems, credit trading allows firms with high pollution abatement costs to do less mitigation by purchasing allowances from relatively clean firms with low abatement costs. Similarly, in regulatory systems credit trading allows firms with high compliance costs to fall short of an emissions or other standard by purchasing credits from firms that exceed the standard.

Distance-based taxes. Taxes that vary directly in proportion to how much a vehicle is driven, for example, on busy roads at peak period.

Downstream policy. An emissions policy imposed at the point at which carbon dioxide emissions are released from stationary sources, primarily from smokestacks at coal plants and other facilities.

Economic costs. The costs of the various ways households and firms respond to a policy (e.g., through conserving energy or using cleaner but more costly fuels). Costs also encompass the impact of a new policy on distortions (e.g., to work effort and capital accumulation) created by the broader fiscal system (these costs can be at least partially offset through recycling of environmental tax revenues).

Efficiency standards. Requirements for the energy efficiency of products, usually electricity-using products such as lighting, household appliances, and space heating and cooling equipment. Similar policies are sometimes applied to vehicles though they are more commonly known as fuel efficiency standards.

Emissions control technology. Technologies used to capture emissions at the point of fuel combustion, thereby preventing their release into the atmosphere. Available technologies can dramatically cut local air pollution emissions from power plants and vehicles with costs that are usually modest relative to environmental benefits.

Emissions factor (or coefficient). The amount of a particular emission (carbon, sulfur dioxide, nitrogen oxides, fine particulate matter) released per unit of fuel combustion. For coal and natural gas, emissions factors are expressed in tons per unit of energy; for motor fuels they are expressed in tons per liter of fuel.

Emissions leakage. A possible increase in emissions in other regions in response to an emissions reduction in one country or region. Leakage could result from the relocation of economic activity, for example, the migration of energy-intensive firms away from countries whose energy prices are increased by climate policy. Alternatively, it could result from price changes, for example, increased demand for fossil fuels in other countries as world fuel prices fall in response to reduced fuel demand in countries taking mitigation actions

Emissions pricing. Policies that put a price on carbon or local air emissions

Emissions trading system or scheme (ETS). A market-based policy to reduce emissions. Covered sources are required to hold allowances for each ton of their emissions or, in an upstream program, embodied emissions content in fuels. The total quantity of allowances is fixed and market trading of allowances establishes a market price for emissions. Auctioning the allowances can provide a valuable source of government revenue.

Energy paradox. The observation that some energy-efficient technologies are not adopted by the market even though they appear to pay for themselves through discounted lifetime energy savings that exceed the upfront investment cost.

Environmental tax shifting. Introducing or increasing an environmental tax and simultaneously lowering other taxes, thus leaving net government revenue unchanged.

Externality. A cost imposed by the actions of individuals or firms on other individuals or firms that the former do not take into account (e.g., when deciding how much fuel to burn or how much to drive).

Feebate. A policy that imposes a fee on firms with emission rates (e.g., carbon dioxide per kWh) above a "pivot point" level and provides a corresponding subsidy for firms with emissions rates below the pivot point. Alternatively, the feebate might be applied to energy consumption rates (e.g., gasoline per kilometer) rather than emissions rates. Feebates are the pricing analog of an emissions or energy standard, but they circumvent the need for credit trading across firms and across periods to contain policy costs.

Fiscal dividend. The revenue gain from energy tax reform.

Flue-gas desulfurization units (scrubbers). Technologies used to remove sulfur dioxide from exhaust flue gases of fossil-fuel power plants. The most common technology is wet scrubbing using a slurry of alkaline sorbent (usually limestone, lime, or seawater). Flue-gas desulfurization can remove 90 percent or more of the sulfur dioxide in the flue gases of coal-fired power plants.

Glossary

Fuel efficiency (or fuel economy) standards. Policies that regulate the allowable fuel use per unit of distance (or distance per unit of fuel use) for new vehicles, often averaged over a manufacturer's vehicle fleet.

Getting (energy) prices right. Reflecting both production costs and environmental damages in energy prices faced by energy users.

Gigajoule (GJ). A metric term used for measuring energy use. For example, 1 GJ is approximately equivalent to the energy available from 278 kWh of electricity, 26 cubic meters of natural gas, or 26 liters of heating oil. One gigajoule is equal to one billion joules.

Gigatonne (Gt). 1 billion (109) tonnes.

Global Burden of Disease (GBD) project. An effort by the World Health Organization to describe the global distribution and causes of a wide array of major diseases, injuries, and health risk factors (including pollution-related illness).

Global positioning system (GPS). A space-based satellite navigation system that can provide information on where and when vehicles are driven.

Global warming. The rise in observed globally averaged temperature from pre-industrial levels that is largely attributed to rising atmospheric accumulations of greenhouse gases (as opposed to other factors like changes in solar radiation).

Greenhouse gas (GHG). A gas in the atmosphere that is transparent to incoming solar radiation but traps and absorbs heat radiated from the earth. Carbon dioxide is the predominant greenhouse gas.

Health or mortality risk value. Value attached to a premature death or other health effect from pollution exposure or traffic accident used to monetize health risks, needed to assess corrective energy taxes.

Integrated Assessment Model. A model that combines a simplified representation of the climate system with a model of the global economy to project the impacts of mitigation policy on future atmospheric greenhouse gas concentrations and temperature.

Interagency Working Group on the Social Cost of Carbon. A group of representatives from U.S. executive branch agencies and offices tasked with developing consistent estimates of the social costs of carbon for use in regulatory analysis.

Intergovernmental Panel on Climate Change (IPCC). The IPCC assesses the scientific, technical, and socioeconomic information relevant for understanding climate change.

Intake fraction. The average pollution inhaled per unit of emissions released, usually expressed as grams of fine particulate matter inhaled per ton of primary emissions.

Kilowatt hour (kWh). A unit of energy equal to 1,000 watt hours or 3.6 million joules.

LandScan data. Provides population counts by grid cell across different countries.

Local or ambient air pollution. Outdoor pollution other than carbon emissions caused by discharges from fossil fuel consumption and other sources.

Megawatt hours (MWh). A unit of energy equal to 1,000,000 watt hours or 3.6 billion joules.

Metric tonne (or ton). A unit of mass equal to 1,000 kilograms, or 2,205 pounds. In the United States, the short ton, equal to 907 kilograms or 2,000 pounds, is more commonly used.

Microgram (mg). Unit of mass equal to one millionth of a gram.

Micrometer (mm). A distance equal to one millionth of a meter.

Millennium Cities Database for Sustainable Transport. Database providing travel speeds and various transportation indicators for 100 cities across many different countries.

Nitrogen oxides (NO_x). Pollutants generated from the combustion of coal, petroleum products, and natural gas. Nitrogen oxides react in the atmosphere to form fine particulates, which are harmful to human health.

Offset. A reduction in emissions in other countries or unregulated sectors that is credited to reduce the tax liability or permit requirements for emissions covered by a formal pricing program.

Ozone. A secondary pollutant formed at ground level from chemical reactions involving nitrogen oxides and volatile organic compounds that can have health effects.

Particulate matter (PM). Particulate matter is classified into fine particulates ($PM_{2.5}$, with diameter up to 2.5 micrometers) and coarse particulates (PM_{10}, with diameter up to 10 micrometers). Fine particulates are damaging to human health because they are small enough to penetrate the lungs and bloodstream, thereby raising the risk of heart, lung, and other diseases.

Parts per million (ppm). Unit for measuring the concentration of greenhouse gas molecules in the atmosphere by volume.

Passenger car equivalent (PCE). The contribution to congestion from a vehicle-kilometer driven by another vehicle (e.g., a truck) relative to congestion per vehicle-kilometer driven by a car.

Pavement damage. Wear and tear on the road network caused by vehicles, especially heavy trucks (because damage is a rapidly escalating function of a vehicle's axle weight).

Pay-as-you-drive (PAYD) auto insurance. A system of car insurance in which motorists pay premiums in direct proportion to the amount they drive.

Petajoule (PJ). A metric term used for measuring energy use. A petajoule is equal to one quadrillion joules.

Primary air pollutant. A pollutant, such as sulfur dioxide or nitrogen oxides, that subsequently transforms through chemical reactions in the atmosphere into a "secondary" pollutant, most importantly, fine particulates, which, in turn, have harmful effects on human health.

Primary energy consumption. The energy content of fossil and other fuels before any transformation into secondary energy (e.g., electricity).

Proxy environmental tax. An environmentally related tax that forgoes some environmental effectiveness because it is not directly targeted at the source of environmental harm. For example, unlike a direct tax on emissions, a tax on electricity consumption does not promote use of cleaner power generation fuels.

Rebound effect. The increase in fuel use (or emissions) resulting from increased use of energy-consuming products following an improvement in energy efficiency, which lowers their operating costs.

Revealed preference. Studies that use observed market behavior to estimate people's trade-offs. For example, estimates of how much people are willing to pay for reduced mortality risk have been inferred from studies that look at the lower wages paid for jobs with lower occupational risks.

Secondary air pollutant. A pollutant (the most important of which from a health perspective is fine particulates) formed from atmospheric reactions involving primary pollutants such as sulfur dioxide and nitrogen oxides.

Secondary energy. An energy source, primarily electricity, produced by combusting a primary fuel.

Social cost. The sum of private cost and external cost.

Social cost of carbon (SCC). The present discounted value of worldwide damage from the future global climate change associated with an additional ton of carbon dioxide emissions.

Stated preference. Studies that use web-based or other questionnaires to infer people's preferences (e.g., their willingness to pay for reducing fatality risk).

Sulfur dioxide (SO_2). A pollutant caused by the combustion of fuel, primarily coal, that reacts in the atmosphere to form fine particulates with potentially harmful effects on human health.

Targeting the right base. Levying charges directly on the sources of an environmental harm to address all opportunities for reducing that harm.

TM5-Fast Scenario Screening Tool (FASST). A simplified model linking pollution emissions from different sources and geographical sites to ambient pollution concentrations and health risks in different regions.

United Nations Framework Convention on Climate Change (UNFCCC). An international environmental treaty produced at the 1992 Earth Summit. The treaty's objective is to stabilize atmospheric greenhouse gas concentrations at a level that would prevent "dangerous interference with the climate system." The treaty itself sets no mandatory emissions limits for individual countries and contains no enforcement mechanisms. Instead, it provides for updates (called protocols) that would set mandatory emissions limits.

Upstream policy. An emissions pricing policy imposed on the supply of fossil fuels before their combustion (e.g., at the refinery gate for petroleum products or the minemouth for coal).

Value of travel time (VOT). The monetary cost that people attach to the time used per unit of travel.

Volatile organic compounds (VOCs). Primary pollutants that are released during motor fuel combustion and react in the atmosphere to form ozone.

Index

Page numbers followed by *b, f* or *t* refer to boxed text, figure captions, or tables, respectively.

A

Air pollution
  from coal, 80–87, 136
  current shortcomings in energy pricing for effects of, 5
  damage per unit of fuel, 4, 39*b*, 87–89
  energy tax design features, 3, 39*b*
  environmental damage from, 20–22
  estimated effects of fuel tax reform, 7–8, 7*f*
  fuel tax design and, 21–22, 127
  harmful components, 16*b*, 68
  indoor, 16–17*b*
  intake fraction calculations, 69–70, 69*b*, 71, 83–85
  international comparison, 21, 21*f*
  mortality related to, 1, 20–21, 22*f*, 67, 73–76, 75*f*, 144, 145*f*
  from motor fuels, 20, 48, 139
  from natural gas, 5, 20, 136
  primary and secondary components, 19–20
  sources of, 19–20
  valuation of damages from, 4, 20, 67–68, 80–87, 82*f*, 84*f*, 85*f*, 86*f*, 87*f*, 90, 92–95*t*
  See also Emissions capture; Population exposure to air pollution; *specific pollutant*
Ammonia, 70, 71, 85
Australia
  air pollution damage valuation, 82
  carbon pricing scheme, 41*b*
  corrective coal tax estimates, 132
  motor fuel taxes, 141
  population distribution, 14
  traffic accident costs, 115
  traffic congestion costs, 108
  See also International comparison

B

Brazil
  air pollution damage valuation, 82
  fuel mix, 13
  motor fuel taxes, 137–139, 141
  potential revenue from corrective fuel taxes, 8, 144
  See also International comparison

C

Canada, 41*b*. *See also* International comparison
Carbon dioxide
  atmospheric concentration and persistence, 17
  carbon charge modeling methodology, 4, 65
  corrective charges for, 5
  current carbon pricing programs, 26
  effects of corrective taxes on emissions of, 7–8, 7*f*, 144–145, 146*f*, 148, 157–160*t*
  emissions patterns and trends, 17, 18*f*
  energy tax design features, 3, 46, 47
  environmental damage per unit of fuel, 89
  global emissions, 17, 18*f*

global per capita emissions, 13, 15*f*
global temperature change and, 1,
  17–18
long range impacts, 66
natural gas power emissions, 5
per kilowatt hour standard, 34–36, 51
significance of, as pollutant, 17
strategies for reducing, 32, 34, 35*f*
See also Air pollution
Carbon monoxide, 16*b*
Carbon pricing programs
  cost-benefit approach to pricing in,
    65–66
  cost-effectiveness approach to pricing
    in, 66–67
  current implementation, 26
  economic costs, 37*b*
  effectiveness of, 34, 35*f*
  effects in energy-intensive industries, 59
  illustrative value, 67
  price determinations, 65
  regulatory policies versus, 34–36
Chile
  air pollution damage valuation, 82
  motor fuel taxes, 141
  population distribution, 14
  potential revenue from corrective fuel
    taxes, 144
  traffic accident costs, 114
  See also International comparison
China
  air pollution damage valuation,
    82, 86
  corrective coal tax estimates, 132,
    134–136
  costs of air pollution, 20
  costs of premature mortality, 1
  fuel mix, 13
  motor fuel taxes, 141
  potential revenue from corrective fuel
    taxes, 144
  potential revenue gains of energy tax
    reform, 8
  traffic congestion costs, 108
  See also International comparison

Coal power
  air pollutants from, 20, 136
  air pollution damage estimates, 80–87,
    84*f*, 85*f*, 86*f*, 87*f*
  consumption data and calculations,
    146–147
  current taxes, 131, 132*f*, 161–164*t*
  current undercharges for effects of, 5,
    145–146
  emissions per unit of fuel, 88–89
  environmental impacts of corrective
    taxes, 157–160*t*
  environmental tax design, 46, 131–136,
    132*f*, 133*f*
  estimated corrective taxes by country,
    149–152*t*
  estimated effects of corrective taxes, 4,
    7–8, 7*f*, 142–146
  estimating population exposure to air
    pollution from, 68–73
  fiscal impacts of corrective taxes,
    153–156*t*
  fuel prices, 131
  health impacts of corrective taxes,
    157–160*t*
  international comparison of emission
    rates, 89, 90*f*
  international comparison of fuel mix,
    13, 14*f*
  potential revenue from corrective taxes,
    143*f*, 144
  smokestack height, 72
  unintended consequences of
    environmental taxes, 45*b*
Compensation payments, 42
Concentration-response function, 74–76
Consumers
  acceptance of energy-saving technologies, 56*b*
  demand reduction among, 32, 33*b*, 34,
    126–127, 147–148
  opposition to energy tax reform, 57–59
Corporate taxation, 40, 59
Corrective taxes
  air pollution levels and, 21–22, 39*b*

balanced design, 38
broader fiscal outcomes of, 45*b*
clean energy technology development and, 54–56
coal, 131–136, 132*f*, 133*f*
components, 3–4
cost effectiveness, 36
country estimates, 149–152*t*
current, by country, 161–164*t*
current revenue from, 24–25, 25*f*
design challenges, 2–3, 8, 165
design principles, 42–43, 60
effect on carbon dioxide emissions, 7*f*, 8, 144–145, 146*f*
effect on fuel consumption, 147–148
emissions trading system versus, 42–44, 50–51
energy prices and, 40, 147
environmental outcomes, 157–160*t*
estimating impacts of, methodology for, 142–143
findings and recommendations, 3, 5–8
fiscal outcomes, 39–42, 153–156*t*
fuel efficiency standards and, 51
health outcomes, 7, 7*f*, 144, 145*f*, 157–160*t*
heating fuels, 47
impact on low-income households, 57–58, 58*f*
interaction with other environmental policies, 50–51
in low-income countries, 60
motor fuels, 47–48, 137–142, 139*f*, 140*f*, 141*f*
natural gas, 136–137, 137*f*, 138*f*
novel alternatives, 52–53
obstacles to implementation, 56–59
offset provisions, 44
other tax reductions offset by, 41–42, 41*b*
potential revenue gains from, 7*f*, 8, 143–144, 145
power generation, 46
problems of current energy pricing, 1
rationale, 1, 2–3, 8, 145–146, 165
regulatory policies and, 34–36, 51–52
traffic congestion and, 23
See also Gasoline tax
Cost-benefit balance
carbon pricing calculations, 65–66
corrective tax calculation, 38
importance of, 38
policy goals, 2
Cost-effective policies, 2, 36–37, 56*b*, 66–67

D

Data sources and quality, 3, 11, 146–147, 165
Demand reduction, 32, 33*b*, 34, 126–127, 147–148
Diesel fuel
air pollution from, 20
corrective tax rationale, 146
current taxes, 26*f*, 61–164*t*, 125–26
damage per unit of fuel, 89
distributional incidence of subsidies, 57, 58*f*
environmental impacts of tax reform, 157–160*t*
estimated corrective taxes by country, 149–152*t*
estimated effects of fuel tax reform, 142–146
fiscal impacts of tax reform, 153–156*t*
health impacts of tax reform, 157–160*t*
potential revenue from corrective taxes, 143*f*, 144
tax design, 6–7, 48, 108, 113, 117, 126–128, 141–142, 141*f*
Discounting costs and benefits of mitigation, 66
Distance-based taxes for vehicles, 48–50, 49–50*b*, 53–54*b*, 101, 146

E

Economic growth
environmental taxation and, 40–41
goals of energy policy reform, 2

Egypt
  fuel mix, 13
  motor fuel taxes, 137–139, 141
  natural gas prices, 136
  potential revenue from corrective fuel taxes, 8, 144
  See also International comparison
Electricity consumption
  per capita international comparison, 11, 12*f*
  strategies for reducing carbon dioxide emissions, 32
  taxes, 26, 36
  See also Coal power
Emissions capture and reduction
  environmental damage per unit of fuel from plants with, 89
  environmental taxes and, 3, 51, 132–133, 133*f*
Emissions trading system
  administrative structure, 44
  advantages of, 34, 44
  cap, 43
  cost effectiveness, 36–37
  design principles, 42–43
  environmental taxes versus, 42–44, 50–51
  European Union experience with, 43, 43*f*
  offset provisions, 44
Employment outcomes of environmental policies, 37*b*, 40–41
Energy efficiency, 32, 34, 36, 44, 48, 50, 51
Energy paradox, 56*b*
Energy prices
  ceilings and floors, 52
  compensation payments, 42
  cost-benefit balance of policies, 38
  current shortcomings of, 1
  energy taxes and, 40, 147
  energy use and, 142
  estimating impacts of corrective energy taxes, 142–143
  goals for fiscal policy reform, 1, 2
  implementation of reforms, 8, 56–59
  opportunities for improving, 27
  policy design challenges, 2–3, 8, 165
Energy security, 17*b*
Energy use
  data sources and calculations, 146–147
  energy-intensive industries, 59
  environmental damage and, 1, 14, 16–17*b*
  fiscal policies to correct side effects of, 1, 2
  global climate change and, 1, 17–18
  harms in extraction and production, 16*b*
  per capita primary consumption, 11, 12*f*
  price elasticities, 142, 147–148
  strategies for reducing carbon dioxide emissions, 32
  unintended consequences of environmental taxes, 45*b*
  See also Coal power; Diesel fuel; Gasoline tax; Natural gas; Subsidies, energy use
Environmental damage
  from air pollution, 20–22
  components of, 4, 44–46
  corrective tax outcomes, 157–160*t*
  cost-benefit approach to mitigation, 38, 65–66
  cost-effectiveness approach to mitigation policies, 66–67
  current fuel taxes and, 25–26
  data quality, 165
  discounting, 66
  energy use and, 1, 14, 16–17*b*
  goals of fiscal policy reform to reduce, 2
  international comparison of fuel mix and, 13
  from larger vehicles, 142
  measurement of, 3
  per unit of fuel, 87–89
  problems of current energy pricing, 1

rationale for fiscal policy reform to address, 31–32
See also Air pollution; Corrective taxes; Global climate change
Equity issues in global mitigation, 67
European Union emissions trading system, 43, 43*f*

F

Feebates, 52–53
Finland, 114. *See also* International comparison
Fiscal policy
  to address externalities, 14
  broader fiscal outcomes of environmental taxes, 39–42
  cost-benefit balance in, 2, 38
  cost effectiveness of, 2, 36
  economic costs of environmental policies, 37–38*b*
  goals, 2
  political context, 4, 31
  principles of environmental protection with, 31–32
  for reducing carbon dioxide emissions, 32
  revenue from corrective taxes, 8, 143–144, 145

G

Gasoline, environmental effects of
  air pollution, 20, 48
  costs of traffic congestion, 102*b*
  damage per unit of fuel, 89
  scope of, 47
Gasoline tax
  carbon component, 139
  competition and, 45*b*
  current, 25–26, 26*f*, 139*f*, 141, 161–164*t*
  design, 47–48, 126–128, 137–141, 139*f*, 140*f*

distributional incidence of subsidies, 57, 58*f*
efficiency and emissions standards and, 51
environmental impacts of corrective, 157–160*t*
estimated corrective taxes by country, 149–152*t*
estimated effects of corrective, 142–146
fiscal impacts of reform, 153–156*t*
fuel efficiency considerations, 128
health impacts of reform, 157–160*t*
potential revenue from corrective, 143*f*, 144
price distortions, 45*b*
rationale, 5–7, 146
Germany
  air pollution damage valuation, 81–82, 83
  energy taxes, 41*b*
  motor fuel taxes, 137–139, 141
  motor vehicle taxes, 50*b*
  potential revenue from corrective fuel taxes, 144
  traffic congestion costs, 108
  See also International comparison
Global Burden of Disease, 83
Global climate change
  carbon charge modeling methodology, 4
  cost-benefit approach to policies to address, 65–66
  cost-effectiveness approach to policies to address, 66–67
  energy use and, 1, 17–18
  estimated effects of corrective energy taxes, 144–145
  See also Temperature rise, global
Greenhouse Gas and Air Pollution Interactions and Synergies, 87, 88*b*
Greenhouse gases
  global temperature change and, 1, 18
  sources of, 17
  See also Carbon dioxide; Nitrogen oxides; Sulfur dioxide

H

Health outcomes
  air pollution, 20–21
  calculation of corrective tax effects on, 146–147
  carbon charge modeling methodology, 4
  concentration-response function, 74–76
  corrective tax effects, 144, 157–160t
  current shortcomings in energy pricing for, 5
  individual susceptibility, 73
  occupational risks in fuel production industries, 16b
  particulate matter emissions, 20–21
  population exposure to air pollution, modeling, 68–73
  traffic accident costs, 112–113, 125–126
  See also Mortality
Heart disease, 74
Heating fuel
  natural gas, 72, 137
  strategies for reducing use of, 32
  tax design for, 47, 60
Heat map of corrective coal tax estimates, 134, 135f

I

Income taxes, 40
India
  air pollution damage valuation, 83
  fuel mix, 13
  motor fuel taxes, 141
  natural gas prices, 136
  population distribution, 14
  potential revenue from corrective fuel taxes, 144
  traffic accident costs, 115
  traffic congestion costs, 108
  See also International comparison
Indonesia
  air pollution damage valuation, 83
  potential revenue from corrective fuel taxes, 8, 144
  See also International comparison
Infant mortality, 74
Insurance, auto, 53–54b
Intake fractions, 69–70, 69b, 71, 83–85
Integrated assessment modeling, 65–66
Intellectual property rights, 55
International comparison
  air pollution, 21, 21f
  air pollution damage estimates, 80–87, 82f, 84f, 85f, 86f, 87f, 92–95t
  air pollution mortality, 21, 22f
  carbon dioxide emissions, 17, 18f
  carbon dioxide emissions, per capita, 13, 15f
  city-level transit delays, 103, 103t
  coal plant emission rates, 89, 90f
  corrective coal tax estimates, 131–136, 132f, 133f
  corrective motor fuel tax estimates, 137–142, 139f, 140f, 141f
  corrective natural gas tax estimates, 136–137, 137f, 138f
  country-level transit delays, 104t
  current fuel excise taxes, 161–164t
  electricity consumption, per capita, 11, 12f
  environmental damage per unit of fuel, 89
  environmental effects of corrective tax, 157–160t
  estimated corrective taxes, 149–152t
  fiscal impacts of tax reform, 153–156t
  fuel energy mix, 13, 14f
  health outcomes of corrective tax, 157–160t
  mortality risk from air pollution, 74, 75f
  mortality risk values, 80, 81f, 81t
  motor vehicle ownership, 11–13, 13f
  motor vehicles and road capacity, 23, 23f
  population density, 13–14, 15f

potential revenue from corrective fuel
   taxes, 143–144, 143f
primary energy use, per capita, 11, 12f
traffic accident costs, 114–115, 115f,
   116f
traffic accident fatalities, 114
traffic congestion costs, 108–109, 109f,
   110f
value of travel time, 106–107, 108f
See also Organization for Economic and
   Community Development
Israel
   environmental damage per unit of fuel,
      89
   motor fuel taxes, 141
   population distribution, 14
   potential revenue from corrective fuel
      taxes, 144
   traffic congestion costs, 108
   See also International comparison

J

Japan
   air pollution damage valuation, 81–82,
      83, 86
   environmental damage per unit of fuel,
      89
   motor fuel taxes, 137–139, 141, 142
   population distribution, 14
   potential revenue from corrective fuel
      taxes, 8, 144
   traffic accident costs, 115
   traffic congestion costs, 108
   See also International comparison

K

Kazakhstan
   air pollution damage valuation, 82
   fuel mix, 13
   motor fuel taxes, 141
   potential revenue from corrective fuel
      taxes, 144

traffic accident costs, 115
traffic congestion costs, 108
See also International comparison
Korea
   air pollution damage valuation, 81–82,
      83
   motor fuel taxes, 137–139, 141, 142
   traffic congestion costs, 108
   See also International comparison

L

Lead exposure, 16b
Low-income countries, 60, 67
Low-income households, 57–58
Lung cancer, 74

M

Methane, 17
Mexico
   air pollution damage valuation, 82
   fuel mix, 13
   potential revenue from corrective fuel
      taxes, 8, 144
   See also International comparison
Millennium Cities Database for
   Sustainable Transport, 102
Mortality
   from air pollution, 1, 20–21, 22f, 67,
      73–76
   calculation of corrective tax effects on,
      148
   carbon charge modeling methodology, 4
   corrective tax outcomes, 157–160t
   economic costs, 1
   estimated effects of corrective energy
      taxes, 7–8, 7f, 144, 145, 145f
   human capital approach to valuing, 77,
      77b
   indoor air pollution, 16–17b
   infant, 74
   international comparison of risk
      valuations, 80, 81f, 81t

from motor vehicle use, 1, 24, 24*f*
risk valuation, 76–87
from traffic accidents, 113–114, 125
value of life saved, 79–80
Motor vehicle use
air pollution from, 20
associated mortality, 1, 24, 24*f*
carbon charge modeling methodology, 4–5
costs of, 1
current fuel efficiency, 128
current fuel taxes, 5–6, 6*f*, 25–26, 26*f*
distance-based taxes, 48–50, 49–50*b*, 53–54*b*, 101
emissions standards, 48, 51, 52
estimated effects of fuel tax reform, 7–8, 7*f*
feebates, 52
fuel efficiency standards, 34, 48, 51
fuel tax design, 3–4, 47–50, 60
international comparison of ownership and, 11–13, 13*f*
recommended corrective tax of fuel for, 5–7
road damage from, 22, 48, 49, 115–117, 127, 142
strategies for reducing carbon dioxide emissions, 32
value of travel time, 106–107, 107*t*, 108*f*
vehicle taxes, 26, 36, 52–53*b*
See also Diesel fuel; Gasoline tax; Traffic accidents; Traffic congestion

N

Natural gas
air pollution from, 5, 20, 72, 136
carbon charge, 136
carbon charge modeling methodology, 4
consumption data and calculations, 146–147
corrective tax estimates, 136–137, 137*f*, 138*f*, 149–152*t*
corrective tax rationale, 146
current price, 136
current taxes, 136, 161–164*t*
environmental damage per unit of fuel, 89
environmental impacts of tax reform, 157–160*t*
estimated effects of corrective taxes, 142–146
findings and recommendations for energy pricing, 5
fiscal impacts of tax reform, 153–156*t*
health impacts of tax reform, 157–160*t*
home heating with, 137
international comparison of fuel mix, 13, 14*f*
population exposure to pollution from, 72
potential revenue from corrective taxes, 143*f*, 144
Nigeria
air pollution damage valuation, 83
fuel mix, 13
potential revenue from corrective fuel taxes, 8, 144
See also International comparison
Nitrogen oxides
atmospheric concentration, 17
damage estimates, 83, 85*f*, 92–95*t*
environmental damage per unit of fuel, 89
natural gas combustion, 136
particulate matter, 20, 68
population exposure, 69, 71, 72–73
share of coal emissions damages, 133–134, 134*f*
sources of, 20, 46, 48
Norway, 49*b*
Nuclear power, 45*b*

O

Offset provisions, 44
Organization for Economic Cooperation and Development, 24–26, 25*f*, 26*f*, 57, 78–79

Outcomes of fossil fuel use
 adverse externalities, 14
 carbon charge modeling methodology, 4–5
 economic costs, 1, 37–38b
 estimated effects of fuel tax reform, 7–8, 7f
 findings, 5
 global climate change, 1
 goals of fiscal policy reform, 2
 mortality, 1
 See also Environmental damage; Health outcomes
Ozone, 20

P

Particulate matter
 damage estimates, 83, 92–95t
 damage per unit of fuel, 89
 estimating mortality risk from, 75–76
 estimating population exposure to, 68–73
 from gasoline combustion, 139
 global concentrations, 21
 health threats from, 20–21, 68
 natural gas combustion, 136
 power generation fuel tax design, 46
 share of coal emissions damages, 133–134, 134f
 sources, 20, 46, 68
Payroll taxes, 40
Poland
 air pollution damage valuation, 81–82
 fuel mix, 13
 motor fuel taxes, 137–139, 141
 potential revenue from corrective fuel taxes, 144
 See also International comparison
Political functioning, 4, 31
Population density, 13–14, 15f
Population exposure to air pollution
 modeling methodology, 68–73
 mortality risk, 73–76, 75f
 valuing mortality risk, 76–87

Power generation, environmental tax design for, 46, 51, 53, 60
Private sector innovation, 54–56

R

Regulatory policies, 34–36, 51–52
Renewable energy
 international comparison of fuel mix, 13, 14f
 policies to promote, energy taxation and, 44, 50, 51
 to reduce carbon dioxide emissions, 32
 subsidies, 34
Research and development, 55
Risk valuation, 76–87
 traffic accidents, 111–115
Road damage costs, 22, 48, 49, 115–117, 118, 127, 142
Russia, 115. *See also* International comparison

S

Shale gas extraction, 45b
Singapore, 49b
Social cost of carbon, 65, 66
South Africa
 air pollution damage valuation, 82, 83
 fuel mix, 13
 motor fuel taxes, 141
 potential revenue from corrective fuel taxes, 144
 traffic congestion costs, 108
 See also International comparison
Subsidies, energy use
 current practice, 26–27, 27f
 data sources, 148
 to firms, 59
 to households, 57–58, 58f
 natural gas, 136
 to promote technology adoption, 56
 for renewable energy, 34

Sulfur dioxide
  from coal combustion, 20, 33b, 46, 82f, 83, 84f, 133–134, 134f
  emissions reduction strategies, 33b
  from gasoline combustion, 139
  international comparison of air pollution damage, 92–95t
  from natural gas combustion, 136
  population exposure to health risks from, 68, 69, 70, 71, 72–73, 86–87
  vehicle emissions, 89
  Sweden, 41b, 50b, 114. See also International comparison

T

Tax policy. See Energy taxes
Tax shifting, 41–42, 41b
Technology development and deployment, 54–56
Temperature rise, global
  current trends, 1
  effects of, 18
  energy use and, 17–18
  projections, 17–18, 19f
Thailand
  air pollution damage valuation, 83
  motor fuel taxes, 141
  population distribution, 14
  See also International comparison
TM5-FASST, 83–87, 96
Trade fees and tariffs, 59
Traffic accidents
  associated risks and costs, 1, 24, 24f, 111–113, 114–115, 115f, 116f, 117–118
  as component of corrective taxes for motor fuel, 5–7, 127, 139–141
  cost calculations, 5, 111, 125–126
  data sources, 111
  distance-based taxes to mitigate costs of, 3–4, 48–49
  external cost assessment, 113–114
  fuel taxes to mitigate costs of, 5–6
  mortality, 24, 24f
  trucks in, 113
Traffic congestion
  city-level delays, 102–104, 103t
  corrective tax rationale, 22–23
  cost calculations, 5, 101, 105–111, 109f, 110f, 117–119
  country-level delay data, 101, 104–105, 117–119, 120–121, 124–125
  distance-based taxes to mitigate costs of, 3–4, 23, 33b, 48, 49–50b, 101
  fuel taxes to mitigate costs of, 5–6, 101, 128, 139–141
  international comparison, 23, 23f
  sources of costs related to, 1, 101, 102b
  strategies for reducing, 33
  vehicle mix as cost factor in, 107–108, 121–123
Turkey
  motor fuel taxes, 137–139, 141, 142
  potential revenue from corrective fuel taxes, 144
  traffic congestion costs, 108
  See also International comparison

U

United Kingdom
  air pollution damage valuation, 81–82
  cost to others of motor vehicle use, 1
  motor fuel taxes, 137–139
  potential revenue from corrective fuel taxes, 144
  traffic accident costs, 114
  traffic congestion costs, 49–50b, 108, 109, 124–125
  See also International comparison
United Nations Framework Convention on Climate Change, 67
United States
  air pollution damage valuation, 81–82, 83, 86
  air pollution mortality estimates, 74–75
  corrective coal tax estimates, 131–132

costs of air pollution, 20
costs of premature mortality, 1
environmental damage per unit of fuel, 89
motor fuel taxes, 141
motor vehicle taxes, 48, 50*b*
per capita primary energy use, 11
potential revenue gains of energy tax reform, 8

traffic accident costs, 114
traffic congestion costs, 108, 109, 124
See also International comparison

V

Value-added tax systems, 40, 40*b*
Value of travel time, 106–107, 107*t*, 108*f*
Volatile organic compounds, 20